PRIMAL
WISDOM
OF THE
ANCIENTS

"An engrossing and informative book, *Primal Wisdom of the Ancients* puts the reader in front of a coruscating kaleidoscope of context with the turn of each page. Rather than research more evidence for cultural diffusion, Laird Scranton takes a refreshing approach from his already opulent opus and successfully amalgamates the purpose underlying the patterns and principles perpetuated by the primordial tradition. This book is not only a valuable contribution to the study of comparative cosmology but also serves as an initiation into the greater mysteries of the esoteric tradition. The only ritual required for the initiate to perform is to read it!"

ANYEXTEE, ESOTERIC RESEARCHER, DOCUMENTARY FILMMAKER,
AND FOUNDER OF ADAPT EXPEDITIONS

PRAISE FOR PREVIOUS BOOKS
BY LAIRD SCRANTON

From *Point of Origin*

"*Point of Origin* is undoubtedly the pinnacle of research into the world's ancient cultures, their mysteries and mythologies . . . a truthful and accurate insight into our origins, encompassing religion, astronomy, mythology, and cosmology. This book is indispensable to anyone seeking answers about our origins."

E. A. JAMES SWAGGER, RADIO HOST AND
AUTHOR OF *THE NEWGRANGE SIRIUS MYSTERY*

"*Point of Origin* is not a book *about* Gobekli Tepe, but it sets that mysterious Anatolian hilltop sanctuary into a matrix of interconnected mysteries from all around the world in a way that is both fascinating and thought-provoking."

GRAHAM HANCOCK,
AUTHOR OF *FINGERPRINTS OF THE GODS*

PRIMAL WISDOM·
OF THE
ΛNCIENTS

The Cosmological Plan
for Humanity

LAIRD SCRANTON

Inner Traditions
Rochester, Vermont

Inner Traditions
One Park Street
Rochester, Vermont 05767
www.InnerTraditions.com

Text stock is SFI certified

Cataloging-in-Publication Data for this title is available from the Library of Congress

ISBN 978-1-64411-028-7 (print)
ISBN 978-1-64411-029-4 (ebook)

Printed and bound in the United States by Lake Book Manufacturing, Inc. The text stock is SFI certified. The Sustainable Forestry Initiative® program promotes sustainable forest management.

10   9   8   7   6   5   4   3   2   1

Text design and layout by Virginia Scott Bowman
This book was typeset in Garamond Premier Pro with Trajan Pro used as the display typeface

Because hyperlinks do not always remain viable, we are no longer including URLs in our resources, notes, or bibliographic entries. Instead, we are providing the name of the website where this information may be found.

To send correspondence to the author of this book, mail a first-class letter to the author c/o Inner Traditions • Bear & Company, One Park Street, Rochester, VT 05767, and we will forward the communication, or contact the author directly at **scrantonlr@aol.com**.

If men learn this [writing], it will implant forgetfulness in their souls; they will cease to exercise memory because they rely on that which is written, calling things to remembrance no longer from within themselves, but by means of external marks. What you have discovered is a recipe not for memory, but for reminder. And it is no true wisdom that you offer your disciples, but only its semblance, for by telling them of many things without teaching them you will make them seem to know much, while for the most part they know nothing, and as men filled, not with wisdom, but with the conceit of wisdom, they will be a burden to their fellows.

PLATO, QUOTING SOCRATES
(FROM *PHAEDRUS*)

CONTENTS

An Introduction
to Comparative
Studies

COSMOLOGY IS THE SCIENCE of the origin and development of the natural structures that exist in our sphere of life. The term *ancient cosmology* refers to the ways in which ancient cultures understood the origins of these same natural structures. In ancient times, cosmological topics were the domain of expert priestesses or priests, who in earliest days were arguably the keepers both of science and of root philosophical perspectives that later gave rise to modern religion. Looked at in retrospect, and knowing humanity's innate capacity for invention, we might well imagine that many distinct systems of cosmology developed from region to region, and these likely framed the processes of creation according to a variety of different conceptions. But after careful comparison of these traditions with each other, what we find instead are surprisingly consistent expressions of what was therefore more sensibly a single archaic system of cosmology. This outlook is testified to by the diverse range of closely aligning concepts, words, symbols, and other elements that characterize these traditions, which go well beyond any reasonable suggestion of coincidence or parallel development. The apparently archaic system that seemingly spawned these traditions was organized around a set of symbols and symbolic themes that Carl Jung would later describe as archetypes, and which

1

persist to this day among the cultures of far-flung societies worldwide.

In modern usage, the word *cosmology* refers to the science of astrophysics, which is the study of how matter and the universe formed. The average modern person is likely to have at least basic knowledge of what an atom is and to know that matter is comprised of atoms and that an electron orbits an atom's nucleus while protons and neutrons are located within that nucleus. They may also have at least some passing awareness of Einstein's theory of relativity and the concept of the big bang and may have heard of more specialized scientific ideas such as string theory and quantum entanglement. For a person's knowledge to go much beyond that would require him or her to have deliberately spent some time reading about or studying subjects that relate to cosmology or astrophysics.

Comparative study of the symbolic cosmologies of various ancient cultures points out many unexpected similarities in their traditions and helps us to clarify the underlying nature of ancient cosmological beliefs. In the strictest sense, the term *ancient cosmology* properly applies to three creational themes: how matter forms, how the universe formed, and how the processes of biological creation occur. In the mind-set of the ancient cosmology, as expressed by the modern-day Dogon tribe of Mali in northwest Africa, these three themes represent parallel processes. Perhaps as a way to underscore that point, in Dogon culture a single progression of symbols serves to simultaneously define all three creational themes. Of course this implies that any given symbol within that progression carries meanings that pertain to all three themes, but with specific nuances of meaning that can vary somewhat in accordance with each theme. Consequently, as we explore the meanings of ancient cosmological stories, symbols, and words it is important that we consider the references in the context of each of the three themes. If, in fact, we are working with a system whose meanings are scientific and whose symbolism was carefully considered, then it makes sense that we most often find the clearest descriptions of cosmological ideas in the earliest representations of ancient cultures, where they may have been expressed through creation myths, as a part of religious hymns or incantations, in inscriptions carved onto the walls of temples, through architectural forms, or in the art of sculpture or paintings whose themes are understood to be mythical or religious. However, among groups such as the Dogon, who chose never to adopt a written language, spoken language becomes the purveyor of symbolic meaning, so we also derive understanding from the phonetics of cosmological terms. We also come to see that symbolism expressing itself through the multiple meanings of ancient cosmological words.

Many of the same cosmological elements are represented in similar form among sometimes widely distant ancient cultures, a circumstance that, if we allow the possibility of deliberate instruction in ancient times by a capable, informed source, strongly suggests that they arose as part of a single earlier tradition. Cultures such as the Dogon and the Buddhists specifically understand their symbolic cosmology as an anciently instructed tradition, and of course one ongoing purpose of our studies has been to test the reasonableness of there being such a common origin. Each favorable comparison argues increasingly against the idea of parallel development or coincidence as the more reasonable conclusion and provides a kind of cross-check on the commonality of the views of the cultures involved, ultimately helping the researcher triangulate on the likeliest original meanings of these elements. Among the surviving traditions that seem to have done the most careful job of preserving original meanings, we count the Sakti (pronounced "Shakti") Cult of Orissa in India, Buddhism in its various forms as they are practiced in India and Asia, and the priestly Dogon tribe of Mali, along with other closely related African tribes such as the Yoruba and the Bambara. Comparative symbols, words, and perspectives of great value to this investigation are also often found in Judaism and in the cosmological viewpoints of Kabbalism. In the context of comparison to Dogon references, we also often find excellent information preserved in ancient Egyptian hieroglyphics, myth, art, and architecture.

Researchers of ancient traditions in India have commented on references that often closely resemble correct cosmological science and so raise reasonable questions about whether this ancient cosmological plan could have originally been scientifically based. With the Dogon tribe, we have a living priestly class who make a specific claim that their cosmology describes how matter forms, a circumstance that elevates this claim to the level of a legitimate interpretation to be explored. In fact, at the most superficial level of comparison it quickly becomes apparent that Dogon descriptions and drawings closely resemble science. Given these circumstances, it is the job of the comparative cosmologist to test the reasonableness of the direct statements made by various cultures regarding the meanings of various esoteric elements with an eye to possible scientific correspondence. At heart, the study of comparative cosmology is primarily a process of corroborating, after the fact, the testimony of multiple witnesses to a common set of formative ideas. Through such comparative studies, the researcher eventually arrives at an overview of a very long-standing ancient tradition, one that seemingly emerged from the mists of a distant archaic era. It is interesting that even the earliest of these traditions seems to have often expressed itself through Jung's archetypes.

We have suggested that the clearest links between symbol and meaning are found with forms that are original to the tradition, and it is a cultural imperative among the Dogon to preserve those original forms. Because of this, the Dogon provide us with a unique window into the archaic era of the cosmology. Language, symbology, and philosophical outlook suggest a path of transmission for the Dogon tradition that began in the Fertile Crescent region and passed down southward and eastward in pre-Vedic times, ultimately through India and into Africa. This inferred history aligns the Dogon cosmological outlook with that of the Samkhya philosophy in India, which is also understood to underlie later religious traditions of India. The root dynamic of material creation as we see it expressed in Samkhya, one that plays out in parallel on all upward levels from microcosm to macrocosm, is a kind of chain reaction that is catalyzed when nonmaterial and material energies come together in the context of an act of perception. If we imagine that these energies arise through a principle of duality as negative and positive aspects of primordial energy, then the underlying dynamic produces the potential effect of a dipole—a pair of equal and oppositely charged or magnetized poles that are separated by a distance. Typically with a dipole, energy is persistently drawn together and then apart again, much like the action of a beating heart. Scientifically speaking, the dynamic begins on the microcosmic level with the interaction of virtual particle pairs that coalesce and disperse much like ripples on the surface of a lake. Any such energies of differing quality that come together tend to form vortices, similar to what we see with whirlpools of water. Electromagnetic energy, when spun, creates angular momentum and evokes Einstein's relativistic mass.

From a broader perspective, according to Dogon cosmological thought, an act of perception causes matter in its wavelike state to draw up like a tent cloth that is pulled upward from its center. The perceived wave begins to vibrate and pivot, and vibration causes mass to encircle and form bubbles, comparable to the dynamic of macrocosmic structures known to astronomers as stellar bubbles. The first few of these bubbles of mass, like soap bubbles blown from a wand, fail to have enough tensile strength to hold their shape, and so soon collapse on themselves. According to the Dogon model, the seventh of these bubbles is ultimately able to hold its shape for a period of time and so, in combination with the previous six collapsed bubbles, produces the first coherent structure of matter. This is known to the Dogon as the egg-of-the-world, or *po pilu*. From one perspective we can think of it as a cluster of collapsed bubbles that come to exist at every point of space-time. Our material universe conforms to four dimensions, and as in Edwin Abbott's classic book *Flatland,* the Dogon understand that the outward appearance of any given figure can differ depending on which dimensional view we take of it. In keeping with that idea, the po pilu is alternately conceptualized as a star with seven rays of increasing length that emerge from a central point, or as the spiral that can be drawn to inscribe the endpoints of those rays.

Over the course of these studies, certain reference books have demonstrated their pertinence again and again, and so take on the status of primary sources. These begin with anthropological studies of the symbolic practices of various groups that have been the focus of our studies. They include Marcel Griaule's *Conversations with Ogotemmeli,* a diary of his thirty-three day initiation into the Dogon esoteric tradition, following years of visits, inquiry, and academic study. Also included is Marcel Griaule and Germaine Dieterlen's finished study of the Dogon religion, *The Pale Fox.* Of similar significance is Francesco Brighenti's *Sakti Cult of Orissa,* which exhaustively documents the practices of a tradition in India that was ancestral to the Hindu and Buddhist religions, and that we also see as ancestral to Dogon culture. Primary information for Buddhist practices is drawn from the works of Adrian Snodgrass, a leading authority on Buddhist architecture and symbolism. These include *The Symbolism of the Stupa; Architecture, Time and Eternity: Studies in the Stellar and Temporal Symbolism of Ancient Buildings*; and *The Matrix and Diamond World Mandalas in Shingon Buddhism.* Supporting evidence from Kabbalism is drawn from works of the philosopher and historian Gershom Scholem. Likewise, since comparative word forms and meanings are often at the root of an interpretation, certain dictionaries also become go-to sources for explanatory information. Geneviève Calame-Griaule's dictionary of the Dogon language, *Dictionnaire Dogon,* is one of these, as is Sir E. A. Wallis Budge's *An Egyptian Hieroglyphic Dictionary.* Frequent reference is also made to Philip M. Parker's *Faroese-English Thesaurus* and Edward Tregear's *The Maori-Polynesian Comparative Dictionary.* Meanwhile, many of the foundational cosmological references of the Dogon symbolic system are intuitively correlated to David W. Thomson III's Aether Physics Model.

From the perspective of our studies, the first evident roots of the tradition were set down in the region of the Fertile Crescent around 10,000 BCE, shortly after the end of the Ice Age. This historical timing places the origins of the tradition at least some seven thousand years prior to the first known written texts, which did not emerge until around 3000 BCE. Based on DNA and linguistic studies, and following the progression of various civilizing skills and religious icons and practices, we can track the influence of this primordial tradition radially outward in all directions from its region of origin in southeast Turkey and western Iran. Signatures of the tradition are seen in Tibet, China, Mongolia, Siberia, and Japan. We find evidence of them eastward from Asia into North America among Native American tribes such as the Navajo, the Cherokee, and the Hopi. To the west we see artifacts of the tradition excavated across Europe as far as Scandinavia and northern Scotland. Perhaps most obviously we trace the tradition to the south and east into India, in tandem with the matriarchal Sakti Cult, where related archaic philosophies and practices are understood to have served as formative influences for the Vedic, Buddhist, and Hindu traditions. Moreover, evidence suggests that the tradition might also be traced even farther to the south and as far eastward as Australia, and to the west as far as ancient Egypt. The many and sometimes subtle commonalities of this symbolic tradition have been the subject of my previous books in this series, whose comparative focus has also ranged geographically from western Africa to Egypt, India, Tibet and China, and ancient Turkey and most recently to such distant locales as the United Kingdom and Polynesia.

The cultures whose cosmological practices we have referred to most frequently treat these archetypes as component elements of an ancient instructed symbolic system. Attending that view is the belief that civilizing skills, closely associated with these same symbols, are said to have been brought to a given culture in ancient times by groups of eight quasi-mythical ancestors, rulers, or deities. In China, the mythical eight are cast as Sage Kings or Virtuous Emperors. In the Dogon culture and Judaism, they are honored as the patriarchs of eight tribal families or lineages. In Egypt, they are treated as eight paired ancestral gods and goddesses of the Ennead or Ogdoad. In the cosmology of the Maori they are described as departmental gods, comparable to Tane, who is said to ascend to the gods and return with baskets of knowledge. By whatever name or classification they may be given, an essential common function is assigned to the eight in each tradition: they are identified as the bringers of a specific set of civilizing skills, arguably the very same skills that make their first appearance in the Fertile Crescent during the earliest era of the symbolic cosmology.

In the Dogon and Buddhist traditions, we are told that a civilizing plan was deliberately tagged to a symbolic cosmology, and that the two were taught in tandem. Often, where we find the archetype symbols, we also see evidence of a common set of civilizing structures, supported by commensurate symbolism. As an example, the traditional ground plan for the earliest civic centers in ancient China takes the same symbolic shape as an Egyptian hieroglyph that Budge identifies as the "town glyph" ⊗. This same shape relates to a traditional Dogon ground plan for agriculture that is expressly cosmologically based, known as the well-field plan. Meanwhile, a comparable plan for agriculture in China, also known as the well-field plan, has been argued by some researchers to have been only theoretical.

Up to this point in our studies, the primary focus has been on demonstrating common elements of the cosmology as they were widely reflected in many ancient traditions. My recent book *Seeking the Primordial* represented a departure from that approach, focusing instead on a consensus of ancient views on how the foundational dynamics of material creation were understood to work and comparing them with those explicitly put forth by the Dogon priests. In that book I took what the Dogon stated in clear language about how a given process was said to work and then compared words, symbols, and concepts from various other ancient traditions to demonstrate a commonality of outlook for each aspect of Dogon thought. More simply put, the goal of the volume was to demonstrate in credible ways that these ancient cultures understood the foundational dynamics of the universe in substantially similar ways. Such synthesis is, to the comparative cosmological elements of these cultures, what a carefully framed inference is to a given set of pertinent facts—we use it to foster a broader perspective by extrapolating an overview from many individual conceptual threads. The idea for this book represents a similar departure from my earlier books: the focus will again be on synthesizing information, rather than primarily gathering it. Here we take the opportunity to discuss what we see as various pedagogical choices that seem to have been made during the formulation of the symbolic cosmology, choices that have bearing both on how we understand ancient symbolism and on our view of it as a designed system. Each of the principles we will be exploring may have its own bearing on key aspects of the tradition that was passed on, so certain themes are likely to come under discussion more than one time, but with different emphasis.

1

MOTIVES AND
INTENTIONS OF THE
ESOTERIC TRADITION

ANY DISCUSSION OF AN ancient and systematized instructed cosmology eventually leads to questions of original intent on the part of those who theoretically introduced it. What possible motives, beyond purest altruism, could have prompted the formulation of such a complex system of symbols and related themes, or justified the immense amount of effort that surely would have been required to introduce it globally to widespread cultures? This is one of the foundational questions that we explored and addressed in my previous book *Seeking the Primordial*. A potential answer to that question rests with the root philosophies of the archaic Samkhya cosmological tradition, where our universe is understood to be paired with a second, sibling universe. The Dogon share Samkhya's outlook that universes form in pairs. The view is that a flow of energy, essential to the life of both universes, scrolls cyclically between them. This energy carries potential mass along with it and so fosters a dynamic in which one universe grows progressively more massive while the other sees a corresponding reduction in mass, somewhat like the effect of sand moving in an hourglass. At the full extent of the cycle, this exchange of energy and mass culminates in what we perceive of as a fully nonmaterial universe and a fully material one. However, Einstein's view of relativity insists that the time frame of an object (or a universe) must slow down as its mass increases, so therefore must also quicken as its mass decreases. Consequently, any virtually nonmaterial universe must persist within a much quicker time frame than the one that we experience. In the context of that quickened time frame, at least from the perspective of an outside observer, all events might seemingly occur at once. We might also see this quickened time frame as comparable to a state of quantum entanglement, where the quantum attributes of two or more particles (such as electrons) can be induced to behave as if they were effectively one particle, without regard to any apparent distance that separates them. An entire universe with its mass minimized to this state of entanglement would outwardly appear as a single unified source, akin to the concept of Unity that defines nonmateriality in many ancient traditions.

The overall cycle of energy that scrolls between the two universes coincides with what the Buddhists refer to as the Yuga Cycle or Great Year. The implication is that, as a by-product of the differing time frames of the two universes, an ongoing shift occurs in humanity's ability to perceive its nonmaterial twin. In Hinduism and Buddhism, the domain of decreasing mass is known as the ascending universe, while the realm where mass increases is known as the descending universe. Another important implication of this energetic flow is that, during the intermediate periods of the cycle, the time frames and relative masses of the two universes must roughly equalize. At this point of parity, it might become thinkable to cross between the universes in much the same way that an airlock makes it possible for humans to transfer between regions of differing pressure. An airlock in a submarine or spacecraft effectively allows us to equalize inside and outside pressures and so facilitates safe movement into or out of the vessel.

When we compare the cycle of scrolling energy metaphorically to that of the Great Year, then this same middle range (the point where parity of mass and relative time frame between the universes would be reached) corresponds conceptually to the equinox of the Great Year, marked out midway between the two metaphoric solstices. In relation to an everyday calendar year, these are the same points that were commonly celebrated by ancient cultures as major holidays, and that continue to correspond to holidays in many modern religions. By both name and symbolic concept, the ancient Egyptians assigned the time of the equinox to Kheper, the dung beetle who represented the concept of *nonexistence coming into existence*. We see one of the equinoxes similarly designated in Judaism by the name Yom Kippur (or if we will allow it, *Day of Kheper*). The spring equinox is also the time of the year when Judaism observes an important holiday known as Passover, which is celebrated with a ritual meal, one that actually ends with the opening of a physical door to the outside, to allow the entry of an imagined nonmaterial spiritual guest named Eliahu. In other words, a Passover celebration seems to enshrine the very effect that we have inferred to be significant through our comparison of ancient traditions.

At the extremes of this same cycle of energy, the fully ascended universe is said to have perfect knowledge (since all events seemingly occur at once) coupled with an inability to act, presumably for lack of any moment of sufficient duration to take effective action. There is a comparable but rare human medical condition, known as locked-in syndrome, where the afflicted individual retains full internal awareness but with no ability to move. While in this condition, it may seem outwardly to others that the person is brain-dead, and so until and unless the affected person can successfully convey the fact of their consciousness to some person around them, it must be a truly horrific state, not unlike being buried alive.

If we grant the qualities of intention and consciousness to each of the twin universes, then the implication is that, somewhere partway into the ascending cycle of decreasing materiality, the ascending universe realizes that the trend will ultimately deliver it to this kind of locked-in state. This realization is expressed in the archaic philosophy of Samkhya as "an idea arises in Purusha that it is bound."[1] Unfortunately, because of the cycle's eroding effect on the ability of the paired material universe to perceive its nonmaterial twin, the less material universe is destined to become locked in during the same epoch that the material universe is least able to perceive it. So the ascending universe becomes motivated, while it is still able, to take positive action on behalf of both universes. The challenge is to establish a structure for society in the material realm that will preserve knowledge in the material universe of the nonmaterial realm's presence during eras when it would otherwise be imperceptible. Put more simply, the nonmaterial is incented, for the sake of its own impending need for an advocate, to essentially help us help it.

For the ascending universe, the goal is to selectively foster a group of sincere, intuitive, informed attendants who understand the broader dynamic of the two universes and who are attuned to the nonmaterial universe's various modes of communication. Each generation of initiates serves as another link in a chain whose ultimate purpose is to foster a group of sincere companions for the nonmaterial universe during the period of its locked-in state. On one level of understanding, the very structure of the esoteric tradition facilitates this selection process. Progress for any given initiate rests with that initiate's own persistence in formulating the next productive question to promote his or her own course of study. If, in the estimation of the initiate's informant, that question is seen as appropriate to the student's own initiated status, then the informant is required by the tradition to respond truthfully—otherwise, the informant is expected to remain silent. This interplay of pertinent question followed by truthful answer continues until the dedicated student is eventually ushered into the inner circles of esoteric knowledge. In Dogon culture, any tribesperson (male or female) may choose to become an initiate, and the dynamics of Dogon society actively encourage the person to make that choice. The women study with female informants and the men with male informants. Moreover, based on the experience of French anthropologist Marcel Griaule it is clear that the option to embark on the journey of initiation is equally open to those from outside of Dogon culture, although such a thing occurs rarely enough that Griaule's informants felt obliged to seek special permission from a board of Dogon priests before actually commencing his instruction.

Looked at in this way, the two universes are treated as siblings who are also each other's companions in an eternal cycle of energetic fluctuation, which cycle Buddhists term the Yuga Cycle. Looked at in this way, the essential purpose of the esoteric tradition is to offset the distancing effects of this cycle by working to secure lasting human memory of the presence of our nonmaterial twin universe. At the same time, the goal is to essentially vet and train a corps of sincere, aware people to act as the material caretakers or companions of the nonmaterial consciousness during the extended era in which it becomes locked in. Because this eternal flow of energy between universes is cyclical and ultimately reverses itself, each of the paired universes takes its alternating turn as both patient and caretaker, and so the relationship between the universes goes forward on a commitment (or perhaps covenant) of mutual support and care.

In order to ensure that this group of caretakers is properly fostered and sustained, the esoteric tradition focuses on two primary instructional goals. The first is to help mankind understand in clear terms what our true relationship is to the larger processes of creation, knowing that we cannot be expected to take correct action without first grasping the true nature of our situation. Absent that, those within the material realm would likely lack the awareness to step up at the proper time to assume the role of cosmic companionship. The second goal of the esoteric tradition, according to the Dogon, is to promote in humanity a facility for discriminating knowledge. This term implies the ability to distinguish truth from error and more essentially assumes an outlook for humanity that rests on good intention. It also implies a practiced ability to draw correct inferences from a given set of facts and relationships. The mastery of such skills among initiates allows certain key details of the cosmological outlook to effectively be "laid between the lines" of what's otherwise overtly stated in the symbolic tradition and thereby serves to misdirect less sincere eyes from certain core cosmological concepts. These goals for humanity seem both sensible and pragmatic, since a proper understanding of the structures of creation, as they are anciently framed, should lead a careful researcher to suspect the presence of an unseen nonmaterial domain. Moreover, precisely because certain creational themes are cast as parallel processes, a person with careful knowledge of esoteric definitions should have the necessary tools to infer its presence.

The dynamic that defines the relationship between initiates of the esoteric tradition and their informants can also ultimately be seen as a symbolic one—it mimics the dynamic of inquiry and response that is said to govern interactions between the two universes themselves. And so, among other things, the typically long period of initiation accustoms the student to the often subtle mode of discourse that is said to take place between the material and nonmaterial domains. According to Samkhya, routine efforts are made from the nonmaterial realm to convey knowledge to (or induce action by) the material realm. These express themselves through the auspices of such effects as vivid dreams, seeming coincidences, the unusual behavior of animals, divination, and clairvoyance. In order to become attuned to these, any material communicant must learn to distinguish between situations that are symbolically meaningful and those that might only give an appearance of being so. Meaningful communication is understood to come with a degree of self-confirmation of meaning, with key ideas or images expressed in more than one way, much as key meanings of the cosmology are often expressed through more than one working metaphor.

As critical to the success of the instructional plan as the dynamics of the esoteric tradition is the need to foster civilizing structures in the material world that have the potential to endure and thereby support the transmission of cosmological knowledge from one generation to the next. Certain choices that may have gone into creating this model are reflected in two highly stable cultural forms, those of the Dogon and of ancient Egypt. We know with certainty that Egyptian culture held its form for some three thousand years, while it seems that Dogon culture, which is arguably based on the same original model, may have sustained itself similarly for upward of five thousand years. In Dogon culture it is not possible to overlook the many ways in which societal practices help to reinforce the instructed cosmology. Griaule and Dieterlen also provide us with insights, which originate with the Dogon priests, into some of the underlying concerns that sparked those choices. Based on those views, the underlying challenge was to establish a stable framework for human society, one based on a cultural outlook that would remain hospitable to cosmological inquiry and discourse. Moreover, for the selection function of the esoteric tradition to work properly, it needed to somehow attract the interest of the most sincere and capable members of the community. We are told that the mythical teachers of the Dogon felt that ways needed to be found to draw humanity to the ideal "like a moth is drawn to a flame." To that end, these teachers seemed to have understood that human curiosity is piqued by things that are hidden or mysterious—that mankind essentially loves a secret—and so one way to continue to draw the attention of intelligent people to cosmology over the long span of ages would be to frame cosmological knowledge as an abiding enigma. We can see how the very nature of multi-ton megalithic stone structures convey this enigmatic effect. The same outlook is reflected in the clustered definitions of the Egyptian word *uha*,[2] which all at once can mean "to solve a riddle," "to separate heaven from earth," and "to hew or cut stone."

A few years ago I participated in a panel discussion alongside my late good friend John Anthony West. He was asked why the ancient Egyptians bothered to build with such cumbersome massive stone blocks rather than with simpler, smaller blocks or bricks. John attributed the choice to the effect that such huge stone blocks have on the psyche of the person who experiences the structure, which is arguably quite different from what would be conveyed by a similar structure constructed from traditional modern bricks. I agree with his point, that as an attracting factor, the sheer sense of awe that a seventy-ton block conveys to those who visit these megalithic edifices cannot be overestimated.

The Dogon priests compare the many levels of symbolism enshrouded in their cosmology to the multiple skins of an onion, which, by the way, is one of the chief products of Dogon agriculture. Metaphorically speaking, again and again an initiate successfully peels back one skin of the onion only to find yet another still waiting to be peeled back. This has also been the essential experience of my studies in comparative cosmology, in that with each book I may seem to touch bottom with the tradition in one way, only to discover that there are other phases of meaning yet to be explored. Again, the effect of that on human curiosity is much the same as a challenging puzzle, which can draw us in as we follow a long trail of related clues to what we hope is a solution.

The Dogon say that a primary goal of the ancient civilizing plan was to lift humanity upward from the status of hunter-gatherers to that of capable farmers. Accordingly, the specific skills said to have been fostered by that plan either bear directly on the founding of an agricultural society or else can reasonably be seen as prerequisites to the establishment of agriculture. The direct relationship between these civilizing skills and the anciently instructed cosmology is most easily seen in the organizational structure of Griaule's very accessible book *Conversations with Ogotemmeli* (originally *Dieu d'Eau,* or "God of Water"), which is presented as a diary of his thirty-three days of instruction as a Dogon initiate. The chapters of the book follow a sequence that reflects the chronology of Griaule's initiation into the Dogon esoteric tradition. Each class of civilizing skill that was handed down is framed as one of eight metaphoric "words," whose definitions are drawn from cosmology. These are set down for the reader in a careful progression. The associated civilizing skills that were passed on include:

- introduction of the concept of clothing
- the concepts of spinning thread and weaving
- the concept of a granary
- the art of classifying plants and animals
- instruction in how to cultivate land
- instruction in fertilization
- instruction in metallurgy, to forge farming implements
- instruction in pottery
- concepts of architecture and the construction of a house
- concepts of fermentation and production of liquor

The broader plan for the practice of agriculture itself rested on premises that were overtly cosmological. Archaically (and perhaps only theoretically) plots of land were originally cultivated in the figure of a spiral, and so associated a rounded shape with the Earth. This is symbolism that is appropriate to the archaic era of the tradition. In actual practice in later eras, land cultivation was defined by a system known as the well-field plan. By this plan, eight square garden plots of land, each measured at eight cubits per side, surrounded a ninth plot, which was used to house a communally shared well. The changeover in form from a spiral plot to a square one is consistent with other symbolic reversals that occurred cross-culturally midway through the millennia. As noted in the introduction, a matching plan is also recognized to have existed in ancient China, at least in theoretical form, where it was again known as the well-field plan. When viewed from a distance, this arrangement of garden plots creates a kind of patchwork appearance on the landscape, one that both the Dogon and the ancient Chinese compared to a quilt.

The arrangement of the Dogon plots, which are still imagined to spiral outward from the central well, repeats the cosmological scheme by which a primordial spiral of matter emerges from a nonmaterial wavelike source in eight progressive stages and so still takes its root in cosmology. The relationship of this cosmological theme to the practice of everyday agriculture exemplifies one of the ways in which cosmology intertwines with structures of daily Dogon life, and so the two aspects of culture can be seen as mutually reinforcing. Once again, that very notion of entwinement can be understood as cosmological. It symbolizes both the relationship of light to mass that is said to exist in the spiral of matter (which the agricultural plot symbolizes), as well as the larger relationship between the nonmaterial and material universes themselves, whose overlap is often characterized as a familial embrace.

It is this same dynamic of interrelated definitions, cosmological and civic, that sets the tone for Griaule's *Conversations with Ogotemmeli,* where each civilizing skill takes its foundation and outward form (wherever possible) from the associated cosmology. Often there are both theoretical and practical presentations of some concepts, where the theoretical form may have been idealized to best reflect the essence of a cosmological idea. As an example of this, the Dogon granary shrine, which we take to represent an archaic counterpart to a Buddhist stupa, is specifically presented as a theoretical construct. In fact, the Dogon priests provide us with a clear set of specifications for how to build their cosmological granary that betray its purely conceptual nature in that they mask an internal contradiction in math, such that if we were to follow the directions to the letter, we would produce a structure that necessarily differs from its explanatory Dogon drawing. The drawing represents a granary with a round base and four flat pyramid-like sides that rise to a square, flat roof of eight cubits per side, which is a numerical match for the defined measure of a Dogon garden plot. However, strict adherence to the plan would result in a stepped structure whose steps rise to a peak, similar to some traditional step pyramids. Within that difference between theory and practical implementation lie many of the variations we typically see in ancient pyramid structures globally. Moreover, the agricultural granaries that are found in Dogon villages take a form that differs markedly from the symbolic granary of Dogon cosmology.

The choice to tag a symbolic cosmology to a practical plan for agriculture can again be seen as a symbolic one. In the cosmological mind-set, symbolism is treated as the natural mode of expression of the nonmaterial, while references that relate to the Earth pertain to the processes of material creation. And so from this perspective, much like the dynamic between a student and his or her informant in the esoteric tradition, the plan of the educational scheme also mimics the embracing interface of the nonmaterial and material domains. To the extent that a physical process such as plowing a field is understood to be symbolic, the very act of plowing reinforces cosmological teaching. Similarly, the simple task of laying out the garden plots in relation to a central well restates knowledge both of the internal structure of the cosmological spiral (which is said to entwine both nonmaterial and material energies) stemming from a water-like source. The framing of these societal and symbolic constructs side by side with each other reminds and encourages an agricultural worker to think deeply about concepts of cosmology while carrying out the often mundane tasks of daily life. Moreover, it's clear that agricultural terms in the Dogon language also reinforce these teachings, since the names assigned to the tools, structures, and processes of agriculture typically derive from root words of the cosmology—terms for the underlying creative processes that are symbolized by the actions. In this way, the structures of daily Dogon life effectively triangulate on and preserve cosmological meanings. Ultimately, the cosmology serves to define the forms and processes of civic life, while those processes, in turn, uphold the structure of the cosmology—again repeating an essential dynamic of the interplay between the nonmaterial and material universes.

In outlining the relationship between cosmology and civilizing skills, the Dogon priests also attest that instruction in the ancient civilizing plan was carried out at a remote location. Given that, it seems sensible that we would find the earliest evidence of many of these skills in the same region and in the approximate era of the Fertile Crescent at the earliest known megalithic sanctuary site, located at Gobekli Tepe in southeastern Turkey. Skills that seemingly made their appearance here included the first cultivation of seeds, the domestication of animals, and the practice of stone masonry and metallurgy. Some researchers cite the likelihood of stellar alignments for the site, which implies that astronomical observations may have been conducted there. Practical factors such as the lack of an evident water source make it clear that Gobekli Tepe was not used as a permanent settlement, it clearly didn't serve as a fortification, nor is a view of it as a burial site justified based on the paucity of human remains found there. The site, which was intentionally covered over within a thousand years of its first use, remained buried until the mid-1990s and preserves a variety of symbolic elements carved in stone. It features a number of circular stone enclosures (many identified with ground-penetrating radar but not yet excavated) that house multi-ton standing-stone pillars with exquisitely carved images of animals, executed both in low and high relief, alongside numerous enigmatic carved symbols. The enclosures also feature stone benches that imply the requirement to seat numbers of people. Taken together, details of the setting, the site itself, the era, and the nearby archaeological finds are all suggestive of an instructional sanctuary. Unlike ancient structures in other regions that are understood to have been rebuilt (some repeatedly) on the same site (in some cases interpreted as the deliberate "decommissioning" of an earlier site), there is no obvious motive or requirement for the building of so many seemingly redundant structures at the same site. The possibility becomes thinkable that these represented projects for a progressive series of classes for stonemasons.

Like any good cosmological term, the two Turkish words *gobekli* and *tepe* can each be translated in multiple ways. Often the meaning of Gobekli Tepe is rendered as "potbelly hill" and so points to a signature icon of the archaic matriarchal Sakti Cult—a clay pot filled with water to represent a womb, also known as a potbelly. Others translate the name as "hill with a navel," which is arguably a symbolic definition of a Buddhist stupa. By yet another rendering, the name *Gobekli Tepe* becomes "central hill," the very same meaning that associates with the Greek suffix *-opolis* that accrued to temple mounds in ancient Greece. Based on that association we can reasonably infer that another intention of the tradition in archaic times was to situate places of instruction at points that were centrally located to the landmasses and to numerous cultures and so reflects a plan that was at least broadly regional, if not global in its conception. In keeping with that outlook, the Gobekli Tepe site sits in a spot that is geographically central to Europe, Asia, the Middle East, India, and Africa. In a later era we see the same to be true with the location of what appears to have been an instructional site on the Orkney Islands in northern Scotland. Orkney is a highly accessible place by sea, made more so by ocean currents of the Atlantic that would have brought a sailing vessel right to its shores.

We have cited likely motivations for the structuring of the esoteric tradition that were symbolic, reflecting the dynamic of interplay between the nonmaterial and material universes, and that seem to have arisen out of mutual self-interest—the fostering of a sincere caretaker for the to-be-locked-in nonmaterial universe. However, there is another potentially credible motivation for having framed the esoteric instructional tradition as a closely held secretive body of knowledge, and this would be to hide its influences from some theoretical third party that might possibly have been hoping to subvert it. Had such a group stood in opposition to these ancient instructional efforts for humanity, shrouding that effort in secrecy would have made it more difficult to thwart. We see signs that such a third party might have made an appearance at around 1500 BCE, during the biblical era of Moses and the Exodus from Egypt, in the form of the god of Moses. Opposition to the more archaic cosmological tradition is suggested generally in the very first commandment to "have no other gods before me" but can also be interpreted more specifically to have been directed against the esoteric cosmology through the mythical plague of the "slaying of the firstborn." In some cultures who share aspects of the archaic tradition, such as the Maori of New Zealand, the term *firstborn* identified candidates to the priesthood. In that context the slaying of the firstborn could represent specific targeting of the priesthood of the archaic religious tradition. The Dogon priests tell us, in keeping with the practices of other cultures, that a key obligation of a Dogon initiate was to feign ignorance about the tradition to anyone who was not a known initiate. So seemingly effective is that precept that a later restudy of Griaule's Dogon work, which characterized the Dogon cosmology as a closely held secret, "found no evidence" of a cosmological tradition comparable to the one that was documented by Griaule and Dieterlen. Meanwhile, the same dynamic of closely held secrecy is also reported by longtime students of other cultures who share significant elements of the same cosmology.

Dogon society reflects an ethic that is consistently nonjudgmental and egalitarian and so reflects an intention to foster those qualities in human society. We see practical expression of this intent in the civic structure of the *toguna*, which is a community discussion house that is a feature of each Dogon village. Whenever a dispute arises, all parties to the dispute are required to retire to the toguna and remain there until the disagreement is settled. This practice helps foster the peaceful atmosphere of a Dogon village. We see that a comparable structure existed as part of the Skara Brae village in the Orkney Islands, where daily life is also understood to have been a broadly peaceful one.

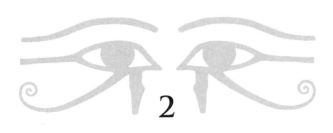

2

WHAT MAKES
US THINK THERE
WAS A PLAN?

THE CHARACTERISTIC THAT MOST immediately differentiates a planned construct from an unplanned one is the degree of organization that we perceive in it. A quintessential comparison might be seen between aerial views of a primordial forest and a cultivated valley. We surely see features of a natural forest that take on an organized aspect, but it would be a very exceptional case for such organization to take on any degree of regularity. Meanwhile, farming almost by definition takes on a regular organization that is outwardly apparent. These same qualities of regularity and organization are reflected time and again in the features of the ancient cosmology.

For any culture that worked strictly from intuition, based on little or no educated perspective as we might modernly think of it, it would make sense to see ritual celebrations that are associated with natural cycles of the sun or moon that are readily observable, such as the monthly phased transition to a darkened new moon. However, the case is that the ancient cultures that shared elements of the symbolic cosmology more often revered ceremonial transitions such as equinoxes and solstices, subtle solar events that are not directly observable and problematic to even predict. It is arguable that ancient cultures might not even have been aware of the timing of these solar events without the help of cosmological tools such as stone circles that gave visibility to them. So from the outset we encounter a set of traditions that rest on someone's preattained understanding that an equinox or a solstice was even a thing. Such understanding implies that someone was already working from an informed perspective regarding the motions of the sun and the moon when establishing these sun-related points of ritual.

We have said that it might seem reasonable to think that the archetypal forms that we observe among widely diffused ancient cultures arose as a consequence of parallel development—the idea that each cultural group, faced with a similar set of societal outlooks and needs, independently came to express these forms in similar ways. However, in many cases, this is not at all how the cultures themselves represent the situation. Comparative studies in cosmology do not properly begin with what we imagine to be true, but rather with what various ancient cultures themselves represented to be true. The fundamental job of the comparative cosmologist is to work from those representations and test the reasonableness of what is flatly claimed by ancient authorities and sources. An excellent place to begin that task is with the Dogon view of their symbolic cosmology as an ancient plan that was transmitted or taught to them, and not one that any ancestral Dogon generation ostensibly developed. By comparison, Buddhism also represents its cosmology as having been an instructed tradition, one that was first passed to humanity in ancient times by a Buddha. The further our understanding of such traditions progresses, the more we see that the outwardly similar forms of many ancient cultures are consistently attended by complex symbolism that argues against parallel development and in favor of a commonly shared archaic source, since no set of specific and discrete esoteric meanings should automatically accrue to any given shape. For example, these ancient cultures do not merely construct shrines, they also take great care to align them—to the four cardinal points, to the equinoxes and solstices, or to specific sets of stars or star groups whose symbolic significance recurs from culture to culture. Moreover, they don't merely align them but do so using a specific progression of geometric shapes that, themselves, symbolize specific things. We can easily understand the common spiritual role that a shrine might play in any religion, but the common choice to build one should evoke no requisite imperative to also align it. So while parallelism of development might reasonably explain the choice to build a shrine, it is inadequate to explain the cluster of symbolic elements that attends to the shrine form in many widespread cultures. In other words, the Dogon and Buddhist perspectives about an instructed tradition arguably align with certain objective facts.

Insight into the likely educational level of those who formulated the symbolic system is also seen in the sheer number of fields of inquiry where mastery of the subject matter is demonstrated. These advanced subject areas include an understanding of the vibrational dynamics of music along with the nature of light and of the various colors of light—both to light itself and to the dynamics of blended colors of light. We see intelligent consideration of the psychology of human motivation and learning. We observe expertise in a wide variety of techniques of material manipulation and construction. We see obvious mastery of, and ability to instruct others in, the skills of agriculture. Someone understood how to domesticate sheep and to progress to the spinning of thread and the weaving of cloth.

If we can infer that these outwardly similar symbolic traditions could not have reasonably arisen independently of one another, then the question becomes one of how the matching elements came to be so effectively distributed to societies across the planet—especially in cases where other evidence of direct interaction between certain cultures has not always been apparent. Perhaps the intuitive outlook is that the symbolic system first took form in one highly influential ancient culture and was then transmitted to each of the others. To that very point, many a nineteenth-century researcher came to tout his or her own pet ancient culture as the originating source of this symbolic tradition, perhaps as a way of enhancing its historical significance. However, in the end, each attempt to trace the tradition forward to the others ultimately met with contradiction in terms of accepted historical time lines and societal circumstances. There simply was no discernable path of cultural transmission that could be traced for the tradition from any single culture to all the others that held up to scrutiny.

Another possibility to explain the widespread commonality of symbolic elements among ancient cultures is the same one previously mentioned, one that we find endorsed by many of the ancient cultures themselves—that of deliberate instruction in ancient times by some capable, informed authority. This view is not often seriously entertained by traditional researchers for the obvious reason of the kinds of intractable obstacles it would seem to present. First of all, what credible candidate could we even nominate to have played the role of such an instructing group in ancient times? Any of the potential answers to that question leads us directly afoul of traditional views of history. The next immediate question may be to ask: How could such a group have managed to communicate with cultures so distantly situated as to not have allowed, in the traditional view, for direct contacts of their own? However, the very nature of a seemingly scientific cosmology forestalls that kind of question because it implies a level of advancement on the part of some active group in ancient times that traditional researchers are understandably reluctant to allow: surely any teacher who could speak with authority on matters of the quantum world would not have seen regional or world travel as any kind of impediment.

Moreover, wouldn't we expect at least some of the ancient cultures to have retained cultural memory of such an advanced group of teachers as well as of the period when the instruction occurred? In regard to this last complaint, in fact culture after culture, including ancient Egypt, actually *does* make specific claims for ancient teachings in subjects ranging from agriculture to weaving to pottery to the art of written language. Many specifically credit the institution of monarchy to godlike influences. More than a few claim actual descent from such a deified group. Ancient instruction is sometimes cast as knowledge that was "gifted" to humanity, a less than specific term that effectively skips past the actual how and why of the getting of it. The most frequently seen mythical paradigm for the imparting of knowledge is one in which a group of quasi-mythical, quasi-deified, or quasi-royal ancestors (most often eight in number) is credited with having "brought" (again an unspecific term) civilizing skills to humanity during the hoary dawn of prehistory, a motif repeated in the lore of ancient societies world-wide. Sometimes, as with the Dogon and in Judaism, these ancestors are celebrated as the patriarchs of revered family lineages. In others, as in ancient China, they are remembered as mythical kings or the founders of ancient dynasties. In yet others, such as in ancient Egypt, specific civilizing skills are associated with or overseen by particular deities, and so in dynastic times the goddess Seshat comes to hold sway over knowledge and measurement at the House of Life and the House of Books associated with each Egyptian temple; Thoth becomes the god of words and language; Khnum comes to be known as a potter god; and Hapy is seen as a god of abundance and the inundation of the Nile. So from the very outset, any sense we may have of a lack of cultural memory for instruction in ancient time by someone more capable than the members of these cultures themselves seems to fly out the window.

Perhaps the most compelling factor that invites us to understand the ancient symbolic tradition as a plan that was handed down lies with the somewhat stunning demonstration of capability and understanding on the part of who may have formulated it. We have noted that the system is understood by the Dogon to reflect three creational themes whose progressions are understood to be parallel: the formation of the universe, the formation of matter, and the processes of biological creation. All three themes are represented simultaneously by a single sequence of symbolic elements, a circumstance that contributes largely to the enigmatic nature of ancient symbolism itself. A similar outlook is reflected in the ancient traditions of India. John A. Grimes, a widely traveled professor of philosophy and religion, writes in his book *Ganapati: Song of the Self*:

> Further, the Vedic tradition itself informs us that three levels of scriptural interpretation happen *simultaneously*, the extrinsic or physical (*ādhibhautika*), the intrinsic or cosmic (*ādhidaivika*), and the transcendent or spiritual (*ādhyātmika*). . . . "It is important to note that these levels of meaning operate simultaneously. Any adequate treatment of the myths' possibilities for interpretation must take them all into account."[1]

When we direct our attention to the somewhat thornier issue of potential candidates for the ancient instructing group, we see that our options are limited from the outset by certain mitigating factors. The first of these pertains to the nature of the symbolic cosmology that entwines with the civilizing plan. The symbolic material is demonstrably scientific, with many of its core meanings recognizable only from the perspective of a society that possesses technology. Moreover, this material is presented as simply a statement of fact, not tentative theory. Our most accessible ancient view into this material rests with the Buddhist concept of *adequate symbols*, shapes whose meanings inhere in forms that occur in nature and so cannot ultimately be lost to the generational chain of initiates of the esoteric tradition. A comparison of Dogon and Buddhist references confirms that such forms include the highly recognizable shape of an electron orbital cloud, the spiral of matter that compares to a Calabi-Yau space in string theory, specific shapes that are produced by the dynamics of energy, and so on. More overt expressions of this same technological basis for the cosmology is found with the Dogon priests, who flatly assert that their symbolic tradition describes how matter forms. Each book in our series of comparative studies has contributed evidence to confirm the objectively scientific nature of

Dogon symbols, concepts, descriptions, and drawings. These references imply a basis of knowledge for the instructional group that might go a bit beyond our own modern scientific knowledge. In other words, the very nature of the material itself implies the influence of some group that must have been quite capable of understanding physics and biology.

Moreover, the choices made in how this scientifically based material is presented within the symbolic cosmology also underscore the very great capability of the instructional group. Again, the simplest view of this lies with the organizational choice to represent three complex creational themes (how the universe forms, how matter forms, and how the processes of biological reproduction happen) through a single progression of carefully framed symbols. It's fair to say that the underlying parallelism of the three themes might not be readily apparent to most modern eyes. We could hardly imagine this feature of the cosmology to have arisen incidentally, since the choice to present it in that way requires an overarching mastery of all three themes. Once again, such mastery implies a group of teachers and transmitters possessed of exceptional capability.

We have said that the inherently scientific nature of the cosmology implies that its meanings were aimed at a technologically capable audience, and therefore a distantly future audience, as seen from an ancient perspective. Consider that we are working with symbolic references that may be twelve thousand or more years old but that arguably reflect concepts that may still today be on the cutting edge of modern science. So a likely concern when formulating these symbols may have been how best to posture them so that their meanings would be evident to that future audience. The ability to convey intuitive meaning depends largely on a person's frame of reference, which is one of the reasons that humor often fails to translate well between cultures or generations—what resonates with the experience of one often fails to do so with another. Moreover, as we have all likely experienced, the intended meaning of a symbol can be an uncertain thing. For example, as citizens of a modern world we learn that a red traffic light means "stop." Future investigators may have to surmise the meanings of green, yellow, and red traffic lights. The simple context of traffic might imply the actions of that traffic—slowing, stopping, and going—as the likeliest meanings for the three colors. If so, the simple sequence of colors should imply that either green or red meant "stop." Meanwhile, a stop sign is also red and actually features the written word *stop* on it, so presuming a mind-set that valued consistent symbolism, cross-comparison of the symbols could help resolve the issue. When we see similar consistency of meaning expressed by ancient symbols, it is again reflective of careful and deliberate thought.

In many cultures, an introduction to the symbolic cosmology is made through a system of esoteric instruction. We have said that in Dogon culture female initiates to the tradition study with women informants, while male initiates study with men. Based on the experiences of French anthropologist Marcel Griaule, who underwent initiation in the Dogon esoteric tradition and carefully documented his process, we know that instruction is carried out according to a rigorous set of rules and guidelines, and that senior priests served as the arbiters of those rules. Structured in this way, the esoteric system again exhibits the attributes of a carefully planned system. We also know that in societies as widespread as ancient Egypt and the Maori of New Zealand, esoteric instruction occurred in formalized schools that were closely linked to temples and the priesthood. The focus of these schools was on a set of subjects that could include agriculture, astronomy, cosmology, stone masonry, zoology, and pharmacology. We know that in the archaic era, which was matriarchal, another focus was on fertility. At the same time, symbolically based organizational features are overtly evident in the structure of aligned ritual shrines, among which the Dogon granary shrine arguably preserves an archaic form. The architecture of the Dogon shrine derives from a set of carefully defined dimensions and produces a theoretical structure with a rounded base, rising to a square flat roof, and culminating in four flat, pyramid-like faces. Central to each face is a ten-step staircase; taken together, these provide a conceptual framework for categorizing various aspects of Dogon life. As an example, initiates are taught to associate constellations of stars with the four faces of the shrine—the associated star groups are used to regulate the annual agricultural cycle. Initiates are also taught to conceptualize ordered hierarchies of plants and animals as "standing" on the ten ascending steps. All of this speaks to a system of knowledge that was carefully considered and organized.

The Dogon granary and Buddhist stupa shrines, which arguably represent alternate forms of the same cosmological structure, also enfold in their symbolism a range of mathematical concepts. We have said that their alignment is accomplished through a geometric method that evokes a matched sequence of shapes. These center on the figures of a circle (symbolic of nonmateriality) and a square (symbolic of materiality). As such, the symbolism of both the base plan of the shrine and the final evoked structure reflect foundational aspects of material creation. The unit of measure on which the geometry is based is a cubit, another feature of instruction that we interpret as one of the "signatures" of the cosmological tradition. Symbolic meaning is established based on relationships between these figures, not on their size, and so a relative unit of measure (as compared to a precise one) served the purpose. A cubit could either be measured as the distance from a person's elbow to the tip of the middle finger or as the length of the average step or pace of a person. Alignment of these shrines begins with a circle drawn around a central point. The Dogon shrine has a radius of 10 cubits and culminates in a flat square roof that measures 8 cubits per side. Mathematically speaking, if we apply an approximate value of 3.2 for pi, the 64-square-cubit area of the roof agrees with the 64-cubit circumference of the circular base and so provides confirmation of intended form for the overall plan of the shrine. The correspondence of these two values again argues for a system of cosmology that involved careful planning.

The language of the cosmology also reflects features of a carefully planned system. Again it is the Dogon whose language perhaps best reflects what we see as the root phonetics of the ancient cosmology. It is traditionally accepted that an oral cosmology preceded the first written language in ancient cultures. It is our view that a number of written glyphs of the Egyptian hieroglyphic language were adopted wholesale from the oral cosmology, where concepts were first implied by phonetic values and often paired with an explanatory drawn image. Any single-syllable Dogon word can be thought of as representing a root phoneme, paired with a root concept. These effectively serve as building blocks that can be mixed and matched in combination with each other to produce more complex words/concepts. Use of many of these same phonemes in languages such as Maori and Hebrew and the hieroglyphic words of ancient Egypt often makes it possible to predict the meaning of a cosmologically related word based solely on its pronunciation. Likewise, we can demonstrate common formulation of certain hieroglyphic words between ancient Egypt and ancient China. All of these aspects of cosmological language uphold an argument for an ancient source of the design for the symbolic system.

3

DYNAMICS AND PRINCIPLES OF SYMBOLISM

SYMBOLISM IS THE ESSENTIAL language of the ancient cosmology—it is the mode by which ancient cosmological information was both conveyed and generationally transmitted. We know that it must be an effective mode of expression precisely because common symbolic meanings can still be broadly demonstrated among so many cultures, millennia later. For the Dogon, who treat the symbol as more essential than the thing it represents, symbolic expression is central to the dynamic interplay of the nonmaterial and material universes—it is the medium by which primordial knowledge is shared. With the work of Freud and Jung, most of us recognize the significance that symbolism holds for human psychology and certainly for dream imagery. There are suggestions that animals such as bats, dolphins, and whales, among others, may communicate with each other through a kind of sonic symbology. Likewise we know that even lack of sight or speech need not be an impediment to a blind or deaf person communicating through symbols. A newborn infant quickly learns to communicate a need through the tone of its cry. Likewise, good parents figure out that consistency in their actions makes it easier for the infant to master aspects of its daily routine. Given all of that, there is a reasonable perspective from which we might take symbolism to be the native mode of conscious expression—one that persists outside of any requirement for spoken or written language.

Perhaps the essential effect of any symbol is a mnemonic one. Much as actors might use stage blocking or their character's physical movements on stage to help recall their spoken lines, so the ancient cosmology associated creational concepts with images, objects, or recurring actions as a way to help an initiate commit them to memory. One of the ultimate goals of the ancient creation tradition was personal mastery of a body of esoteric knowledge, and the use of symbolic devices helped to facilitate that. The benefits for memory of symbolic association increases with repetition, so actors tend to frequently rehearse their spoken lines. In the esoteric tradition, similar repetition is accomplished by assigning symbolism to everyday objects and routine actions so that the rhythms of daily life serve to reinforce the associated meanings.

A symbol serves as a repository of meaning; its essential function is to preserve knowledge in a way that it can later be correctly recalled. That function of recall works best for any given symbol when we see an intuitive relationship between the idea and the symbol that represents it, one that instinctively conjures a particular meaning for us. As an example, it seems only natural to associate the concept of day with the sun. Overall, the function of recall works best for a symbolic system when there is consistency to the methods by which meanings are assigned to symbols.

From the Dogon perspective, the essence of a symbol (i.e., its ultimate meaning) rests with and is retained by the nonmaterial domain. We have already remarked that symbolism is a form of expression that lends itself well to situations in which a person is unable to actually speak. Moreover, one of the widely familiar metaphors of the cosmology specifically associates the processes of material creation with the formation of a Word. So from the outset the dynamic of cosmological expression, which we know is rooted in symbols, turns on an idea or meaning that is formulated nonmaterially and then expressed materially, much as we speak a word. From that viewpoint, we have cause, much as Freud and Jung did, to think of symbolism as a kind of language of outward expression employed by a nonmaterial consciousness, one that is unable to speak its words directly.

The Dogon tell us that the nonmaterial universe is perfectly able to perceive its material twin, but that as the cycle of scrolling energy skews the relative time frames of the universes, the material side becomes progressively less able to perceive its nonmaterial twin. Ultimately the nonmaterial arrives at a state in which it is said to have "perfect knowledge but an inability to act," while the material universe retains "imperfect knowledge, but with full ability to act." We have said that the Samkhya philosophy, upon which the cosmology rests, attests that because of the inhibitions to action in the nonmaterial domain, attempts are routinely made to communicate knowledge to, or to induce action by, the material realm. We pointed out in *Seeking the Primordial* that the esoteric tradition itself is predicated on the dynamic of the two universes, which actually becomes the model for interaction between a priestly informant and an initiate. This relationship is perhaps best illustrated in cultures such as the Dogon and the Maori, where a student who sincerely pursues the esoteric mysteries ultimately fosters the truthful responses from his or her instructor that allow them to progress. Two meanings of the word *dogon*, which are "to complete the words" and "to feign ignorance," reflect obligations of an initiate in the esoteric tradition, or perhaps more precisely the obligations of an initiate and that initiate's informant. Two comparable obligations similarly align with an ancient name for Egypt, Khem, and so imply the influence of the same esoteric mind-set. With this instructional dynamic as our entry point to the tradition, we begin to understand how other key aspects of ancient cosmology may also rest on physical aspects of interaction between the two universes.

In *Seeking the Primordial* we argued that the root dynamics by which material creation is evoked also play out in parallel on all upward scales of creation, beginning in the quantum world and extending conceptually upward to the macrocosm. It is this same essential message that we take away from the Hermetic cosmological theme, "As above, so below," a statement that provides us with a very generous hint about parallel dynamics and how the underlying processes of creation are understood to work. We also know that sacred geometry, as it is defined in the base plan of a Buddhist stupa, was not meant merely to symbolize how space is evoked, but also to illustrate how the geometry of the two universes actually comes to be configured. What emerges as the focal point of this geometry is the shape of the fish or *vesica piscis,* which is defined by a region of overlap between two circles that symbolize the nonmaterial and material realms. This is the same essential shape that is evoked by the action of virtual particle pairs, which seem to fluctuate into and out of existence against the backdrop of a vast cosmic sea of energy. The effect produced by these fluctuations is comparable to the ripples we see on the surface of water, which randomly emanate and dissipate along the interface between a body of water and the domain of air just above it. In that context we can say that key aspects of cosmological symbolism seem to be rooted in the energetic dynamics of the two universes themselves, thus implying that any discussion of this symbolism should properly begin with those dynamics. Such symbolism is easy to recognize because we see it expressed repeatedly in the outward motifs and themes of the cosmology.

The Dogon say that their cosmology rests on principles of duality and the pairing of opposites, and much of the symbology of the tradition reflects these principles. Both are attributes that the two universes embody as they are described in the tradition. Universes represent grand structures of the cosmology and are among those that we are told emerge as pairs, one characterized as feminine and the other as masculine. The same attributes take on many different outward forms in the ancient cultures we have studied, ranging from concepts of yin and yang and the primordial deities Fu-Xi and Nu-Wa in China to male/female deity pairings such as Siva and Sati in India—even in fundamental oppositions that characterize the symbology, such as the notion of separating Earth and sky. As pervasive themes of the cosmology, such principles ultimately lead the thoughtful disciple of the tradition to infer the influence of these original pairings in situations where they might not be overtly stated, as is the case with the highly secretive concept of a female counterpart to Ganesha. Because of the effects of parallel symbolism, the concept of a female Ganesha would imply a nonmaterial counterpart to the Dogon spiral of matter, the po pilu—the inward-turning spindle of what the Kabbalists define as a primordial scroll.

Perhaps the overarching motif of the ancient creation tradition is that of reconciling the nonmaterial and the material. This theme echoes one of the stated goals of the Dogon esoteric tradition, which is to foster in humanity an understanding of our true relationship to the larger processes of creation. The outlook of the Samkhya philosophy is that structures of material creation are evoked when nonmaterial and material energies come together in the context of an act of perception and so implies that any true understanding of these processes may actually rest with our ability to reconcile the comparable attributes of these two domains. It is in the context of this larger relationship that we come to understand the symbolism, as it is framed in the ancient tradition, as sensible.

The geometric figure of a circle is perhaps the first coherent shape that energy assumes in nature. It also characterizes the dynamic of angular momentum and is ultimately responsible for evoking mass. Partly because of this, the figure of a circle comes to symbolize the nonmaterial domain. Meanwhile, our material universe most obviously takes its definition from the emergence of space and so similarly comes to be represented by the geometric figure of a square, which also intuitively conveys the notion of "a space"—perhaps even more so in the three-dimensional form of a cube. Consequently, any cosmological reference that is predicated on the figures of a circle and a square can be understood as a metaphor for the act of reconciling the two universes. Perhaps the most obvious of these metaphors is reflected in the notion of squaring a circle, which again is a dominant cosmological theme in many ancient cultures. A noted feature of ancient Egyptian mathematics was a technique for configuring a square and circle of equivalent areas. The Egyptians knew that a square of eight units per side (configured both like a traditional Dogon agricultural plot and the square roof of their granary shrine) encloses the same approximate area as a circle with a diameter of nine of the same units. This configuration is comparable to Budge's previously discussed Egyptian town glyph ⊗, and it constitutes one of the points of correlation between the ancient Chinese and Egyptian cosmological traditions. Likewise, my friend Ed Nightingale's reverse-engineered plan for structures on the Giza plateau works its effect based on an eight-unit square and a nine-unit circle, configured together in reference to a mathematical grid based on the Platonic numbers—two sequences that arise respectively from the repeated doubling and tripling of the integer 1. A wealth of compelling implications of this plan are set out in his book *The Giza Template*. The plan correctly predicts the arrangement, placement, and dimensions of various structural elements situated at Giza, while also evoking a host of other significant scientific measures and values.

In the mind-set of the archaic cosmology, shrines, sanctuaries, and temples are conceptualized as localities where the nonmaterial and material come together in reconciliation with each other. It is for this reason that the figure of a dome or arch (figures that actually combine a squared shape with a circular one) comes to be associated with these structures. In the early days in the region of the Fertile Crescent, three domes or arches constituted an identifying icon of *chaitya* sanctuaries. The term *chaitya* combines the cosmological phonemes *het/get/chet* ("house" or "temple") with the word *yah*, meaning "light." Our first evidence of three dome-like figures as likely identifying marks for a sanctuary or temple is found at Gobekli Tepe in southeastern Turkey, represented by the three "handbag" figures inscribed on a pillar. These are formulated with a square base and a rounded top, again in keeping with the figure of a hemisphere. Historical expressions of the concept of a shrine again mimic the parallelism of scale that we observe with other structures of the universe. We find almost identical symbolism attached to a chuppah canopy, beneath which a Jewish couple is traditionally married; in the four classic forms of a Mongolian or Siberian yurt, in a portable ancient Egyptian shrine called a *seh;* in the form of a Buddhist stupa; and even in a traditional Native American tepee. This symbolism arguably carries forward to many of the grand cathedrals known to history.

In the consensus mind-set of the cosmology, the vesica piscis, in its role as the *aether unit* gateway, serves to measure out and configure the primordial energy that scrolls between the universes and provides the basis for understanding a Jewish Torah scroll as a symbolic element of the cosmology. The written texts found on the parchment of a Torah scroll are similarly measured out in daily portions, and over the course of a year the scroll shifts to the limits of its extent from one spindle to the other and so mimics the action of scrolling energy and mass between the two universes. The concept, previously mentioned, is referred to in Kabbalism as the primordial scroll. The pointer tool that is traditionally used by the Torah's reader to follow the text of the scroll is known as a yad, and it is symbolic of the forearm and finger that defines a cubit as a cosmological unit of measure. The act of reading a Torah in daily measured portions illustrates both the concept of the flow of time in measured moments and the discrete imparting of esoteric knowledge. Meanwhile, the concept of "knowledge" itself also constitutes one of the cosmological metaphors for a material effect of the scrolling energy.

Even significant astronomical events that are widely celebrated by ancient traditions can be understood as symbolic of reconciling the nonmaterial and material domains. An obvious example is seen in the concept of the heliacal rising of Sirius, which the Dogon represent as the act of Sirius meeting the sun. Because the interaction of the two Sirius stars parallels those of virtual particles, both taken to represent conceptual gateways to the nonmaterial, Sirius becomes a symbol of the nonmaterial universe. Similarly, our sun, which occupies the central point of the orbit of the Earth (like a cosmic sun glyph $\odot$), comes to be an icon of the material universe. So in midsummer when Sirius emerges from a period of invisibility masked by the glare of the sun and rises just ahead of the sun at dawn, the annual event is interpreted as a symbolic reconciliation of the two universes.

The next symbolic motif that clearly pertains to the dynamics of the two universes is that of movement from Unity to Multiplicity, which we also see as a metaphor for how energy expresses itself between them. This theme is rooted in an inherent translational dynamic that occurs at the boundary between the nonmaterial and material realms, one that induces multiple material effects to be evoked by any non-material impulse. In Buddhism, the same concept is illustrated in regard to white light (symbolic of the nonmaterial realm), which is said to enfold within itself all of the colors of visible light. We know, scientifi-cally speaking, that when white light is passed through a crystal, it can evoke a familiar rainbow of seven colors—red, orange, yellow, green, blue, indigo, and violet. If we somehow had no knowledge of the white light that initially passed into the crystal, we might yet infer its pres-ence based on the range of colors that we observe to emerge from it. In other words, there are situations in which the quality of a potentially nonmaterial influence can be inferred from the material effects that it evokes. The circumstances of the rainbow of light, comparable to the seven-note musical scale that arises from the vibration of sound, suggest that the evocation of Multiplicity is an inherent feature of the dynamic of the universes.

This same basic dynamic of one to many repeats for cosmological concepts or words, which typically express themselves in ancient lan-guages as clusters of discrete meanings. We can infer that these meanings properly belong as collective groupings because we find them associated with each other again and again in the languages of widespread ancient cultures. Properly speaking, the clustered meanings "float" with cosmological concepts rather than with strict phonetics, since we often find the same meanings grouped together, even in languages that can sometimes differ markedly from the Dogon or Egyptian models. This same circumstance of clustered meanings often allows us to positively correlate cosmological concepts expressed in Sanskrit with those of the ancient Egyptian and Dogon languages. The relationship of single concept to multiple meanings upholds the outlook previously expressed that the concept itself originates with the nonmaterial domain. In the Dogon language, where words are spoken but not written, these meanings most often express themselves as multiple definitions of the same spoken word. In the Egyptian hieroglyphic language, where the spelling of a word is arguably conceptual and so words that may be pronounced alike may be spelled differently, the clustered meanings play out as homonyms. These clusters greatly enhance our ability to correlate terms between cultures since they provide multiple points of evidence on which to posture a correlation.

The notion of Multiplicity arising from Unity plays a further role in early written language and becomes a defining feature of the Egyptian hieroglyphic language. While with alphabetic languages we expect to see a one-to-one relationship between phonetic values and letters, the opposite is true for Egyptian hieroglyphs, where some four thousand glyphs are utilized to represent approximately forty phonetic sounds. This means that multiple glyphs can often convey essentially the same sound. These glyphs, grouped by their phonetic values, often bear an apparent relationship to the same clustered meanings that we say reflect this effect of cross-universe translation. For example, we can see that glyphs pronounced like the English *n* are associated with images whose names align with our clustered meanings. The word *nu*, which can refer to waves or water begins with a wave glyph 〰〰, while a hieroglyphic word for "pot," *nubti,* begins with the image of a clay pot ○, both assigned a phonetic value of *n.* So, just as the Dogon attest that the "seeds" of material creation are retained by the nonmaterial, there is a perspective from which these glyph/phonetic relationships might also take their root in nonmaterial/material translation.

In this same context we see further ways in which Multiplicity may be an inherent product of an underlying dynamic between the two universes. First, we often see the very same clustered meanings represented in vivid dreams, where other nonmaterial effects may also be evident. Such effects routinely include the dreamer's difficulty in recalling certain aspects of a dream once they have awakened, which we take as another expression of the one-way view (nonmaterial to material) that persists between the two universes—things that essentially transpire nonmaterially are effectively blocked from material perception. A similar effect (assigned by psychiatrists to the subconscious) holds true for other experiences that, because they go against our everyday experience, may also take on dreamlike aspects, such as the memory of a car accident that plays out for us in slow motion or a traumatic incident whose memory may be suppressed and has become blocked. We know that many of the unrecalled details of these dreamlike episodes must persist outside of everyday consciousness because they can often be successfully recalled to conscious memory through hypnotism. Moreover, we often see these same types of clustered meanings and memory effects in eyewitness accounts of UFO encounters, where a craft or a being is credited with a specific action that coincides with a strong impulse, emotion, action or visual effect. Such pairings, which demonstrably recur again and again in UFO encounters, very often play out according to the same clusters of meanings that we observe in ancient cosmology.

The circumstance of one-way viewing between the universes, which is a feature of cosmology in numerous cultures, is itself represented symbolically in the ancient cosmology. This is reflected in rituals of Dogon masks, where the wearer of the mask observes others but cannot themselves be observed. The outlook of the Yuga Cycle is that ever-greater differences of time frame between the two universes eventually erode the material universe's ability to perceive the nonmaterial. An intuitive metaphor might be to a police interrogation room where those of us outside look in, but the suspect within ostensibly cannot see out. It seems reasonable to think that the one-way viewing effect for universes might be a consequence of the unidirectional arrow of time that the nonmaterial universe experiences. The same concept of one-way viewing plays a role in the ascending stages of material creation, most notably in the formation of the seven wrapped dimensional bubbles that comprise the Dogon egg-of-the-world or po pilu, which is the Calabi-Yau space of string theory. It is seen in the Egyptian concept of an Arit, a term based on the cosmological phoneme *ar*, which implies ascension. The concept of an Arit takes on aspects of a gateway, in that Budge tells us that each Arit was attended by a gatekeeper, a watcher, and a herald. The very notion of a watcher, which is again a feature of cosmology found in numerous cultures, is rooted in the idea of a one-way view of the material realm from the nonmaterial realm.

The perspective of a one-way view is also arguably a feature of symbolic written languages such as Egyptian hieroglyphics and Hebrew, where noncacophonous sounds (the vowel sounds) are routinely omitted from written words. We could say that the arrangement of unwritten and written sounds symbolizes an embrace between nonmaterial (unwritten) and material (written) elements. This produces an effect for the symbolic written word that implies a degree of inherent uncertainty in the text. The esoteric message is that the reader who views a written message is only seeing a piece of a total picture and so is required to ultimately infer the unwritten elements that are missing. The concept is again reflective of another of the stated goals of the esoteric tradition, which is to foster in humanity a facility for discriminating knowledge.

Symbolism, both as it functions within the oral tradition and as it suggestively pertains to animal images and icons at places such as

Gobekli Tepe, can be seen as a precursor to written language. The effect of such symbolism is direct, in that the meanings associated with an image or a phoneme are largely direct ones that we infer based on the symbolic meanings of ancient Egyptian words. A more nuanced representation of symbolic meaning is made possible through what we see as the mechanism of defining words in the Egyptian hieroglyphic language, words whose symbols seem to assign meaning to an unpronounced trailing glyph. With many glyphs we discover more than one such defining word in Budge's dictionary, with each seeming to address the symbolism of the glyph from its own perspective. So for example, if our purpose is to represent the concept of the sun, we might theoretically attach the sun glyph to a word for "brightness," a word for "center point," and a word for "heat." Taken together, the alternate definitions offer us a more dimensional outlook on the concept that the glyph was meant to represent. This outlook on the nature of the Egyptian hieroglyphic language also potentially makes sense of the ancient requirement for so many thousands of written glyphs.

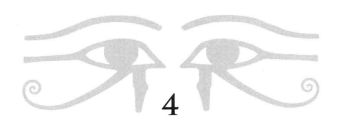

4

METAPHORS OF
THE COSMOLOGY

WE HAVE SAID THAT the scientific images and concepts that populate the ancient cosmology imply that its meanings were aimed at a future audience, one that would have necessarily gained the technological capability to recognize and confirm its references. The many intervening generations, for whom technology might not be sufficiently advanced to allow for that, were charged with the caretaking, preservation, and transmission of those references (not a simple task, to anyone who ever played the children's game of telephone). One challenge for the authors of the symbolic cosmology would have been to couch those references in such a way as to have the best chance of being correctly understood hundreds or thousands of years later. A technique that lent itself to this requirement was that of a metaphor, by which we frame a concept we know a person might not understand in relation to one that is more familiar. The Dogon understand that the act of teaching a person is not always a simple task because it can meet with resistance. In the Dogon view, efforts to convey knowledge often require us to "tug against a thing that pulls back"—in their mind, like pulling on wild grass. It is in this sense that the Dogon people self-deprecatingly refer to themselves as "wild grass," acknowledging that it may simply be human nature to resist instructed knowledge. Even for a modern person the essential experience of arriving at some new bit of cosmological knowledge is often a kind of bemusement over how very dense we were about grasping some seemingly obscure concept that now seems perfectly evident to us.

Precisely because of this seemingly natural human resistance to learning, one of the likely imperatives for those who framed the ancient symbolic system was to find ways to effectively navigate around this resistance. One approach would be to prefer the most direct path to an intuitive understanding of any given concept, and thereby bypass the kind of extended lectures that many of us learned to tune out during our own educational experiences. Of course, as any lecturer who has made a PowerPoint presentation knows, illustrating a concept with a related image is one effective way to do that. Using a short story line to frame a point is another. Meanwhile, insightful comparisons made through a well-selected simile or metaphor can offer a third way to intuitively convey an idea to a person.

Metaphors are shortcuts to new understanding in that they exploit the benefit of things we already know. For that reason, a metaphor is most effective when it resonates with the experience of the person for whom it is devised. A well-chosen metaphor has the potential to convey a complex idea in a direct and insightful way, simply by positing that one thing can be understood to be intimately like another. With the best metaphors, understanding can be immediate and intuitive. However, if the comparison that provides the foundational basis for the metaphor ultimately falls outside of the experience of the person for whom it is intended, then the knowledge to be conveyed will also ultimately miss its mark. So skill at formulating and expressing ideas through a careful choice of metaphors becomes one of the hallmarks of an excellent teacher. In the case of ancient teachers who knew they might well be distanced from their intended audience by thousands of years, selecting the proper metaphors may have seemed problematic, since there could be no certainty as to which ones would be likeliest to resonate with some imagined future recipient.

We see two aspects of the symbolic tradition as having likely been motivated by this problem. First, metaphors of the tradition were commonly rooted in elements of the natural environment that are not prone to change quickly, and that therefore should likely continue to be experienced by people in much the same way over a very long span of time. And so the natural effect of wind, which should constitute a universal effect, became a surrogate for the concept of vibration, while water references, which are also universal, convey the notion of ripples and of waves. The idea of raindrops from a cloud familiarly illustrates the concept of Multiplicity arising from Unity. As is more directly demonstrated by the previously mentioned Buddhist concept of the adequate symbol, elements that inhere in nature can be counted on to remain constant in the human experience, and so provide an excellent basis for cross-generational metaphors. Second, many of the meanings of the cosmology are expressed redundantly in relation to more than just a single metaphor, both as a way to confirm an intentional meaning and perhaps in the hope that at least one of the chosen metaphors might resonate better than another with an uncertain future audience.

The concept of a metaphor itself is also leveraged within the symbolic cosmology, perhaps in keeping with the broad range of symbols that the system comes to encompass. The system defines a range of four-stage metaphors, modeled after the same pattern as the familiar primordial elements of water, fire, wind, and earth. These staged metaphors are intuitive in that many of them reflect progressions that can be readily observed in nature, such as the growth of a plant from a seed or the evolution of a mature bird from an egg. That approach to symbolic expression accustoms an initiate both to the technique of conceptualizing a complex process in stages and to resolving complex issues by breaking them down into smaller, more manageable units. If we take these four-stage metaphors as effective categories for the multiplicity of symbols themselves, then the metaphors make it possible for us to place a related set of progressive symbols into proper sequence with one another. For example, the sequence of water, fire, wind, and earth allows us to infer that any cosmological reference to "waves" must correspond to a process that falls at the early formative end of the scale of material creation, while those to "earth" properly belong at the culminating end.

We are introduced to the notion of four-stage metaphors through a Dogon example that relates to the act of building a structure, where each of the four stages is assigned a more generic cosmological term. The first stage, known in the Dogon language as *bummo*, essentially stakes out the perimeter of a planned structure in geometric points. With the second stage, called *yala*, the plan gains definition through the plotting of additional points (essentially endpoints of geometric lines) from which a more refined outline for the structure takes focus. The third stage, given as *tonu*, fleshes out the structural outline with actual lines that detail the major features of the structure, while the fourth and final stage, *toymu*, refers to the structure in its completed form. This generic four-stage grouping, which serves to define the concept of the other related metaphors, reflects what are ultimately both dimensional and geometric perspectives. We see this in that the stages progress from a geometric point (*bu* means "place") to the concept of a line (*ya/yah* refers to "light," which takes the form of linear rays), then to the notion of an enclosed area (*nu* refers to "water," whose surface is a plane), and finally to a three-dimensional space (*tem* means "complete"). The Dogon names for the four stages align with the comparable ancient Egyptian terms *bu maa, ahau, tennu,* and *temau.* Similarly, whenever the Dogon refer to a cosmological drawing to support their definition of a concept, each drawing is identified with one of the four terms and so suggests that it presents a view that is given from one of these four dimensional perspectives. Based on that outlook, we are explicitly told that the Dogon drawing of the egg-of-the-world as a seven-rayed star represents the yala view, while the interpretation of it as a spiral, drawn to inscribe the endpoints of those rays, represents the tonu view. The intriguing suggestion in this case is that we have be presented with two separate dimensional views of the same structure, a circumstance that offers us potential insights into the nature of dimensionality. In cases where no term is assigned to a Dogon drawing, the implied view is of a stage of completion (toymu), which would also be our own material view.

Perhaps the first of these four-stage metaphors—the familiar metaphoric progression of water, fire, wind, and earth—can be understood on more than one symbolic level. Outwardly these might represent four states of matter: liquid, plasma, gaseous, and solid. However, as the Dogon represent them, the terms also hold meaning in the context of progressive stages of material creation. Water corresponds to matter in its underlying wavelike state, fire to an act of perception, wind to the concept of vibration, and earth to mass. Similarly, another of the group metaphors is given in relation to stages in the growth of a plant. This progression begins with the concept of a seed, then a shoot, then a growing plant, and finally a full-grown plant such as a tree. This metaphoric theme calls to mind the widespread symbolism of a Tree of the World or Tree of Life that is preserved in many ancient cultures. Perhaps more immediately useful to students of ancient Egyptian cosmology is yet another of these group metaphors that takes its definition from conceptual classes of the animal kingdom. Here the progression begins with the category of insects, then fish, then four-legged animals, and ultimately birds. Familiarity with this group metaphor empowers us to contextualize various Dogon animal symbols and Egyptian animal-headed deities and to effectively place their symbolism along a continuum of the stages of creation. In keeping with this metaphor, the dung beetle Kheper (an insect) symbolizes the concept of nonexistence coming into existence; the Dogon Nummo fish drawing (a fish, obviously) reflects the early stirrings of matter; Egyptian and Dogon jackals (four-legged animals) oversee an underworld or second world where the perfect wavelike order of matter is disrupted and fundamentally reconfigured; and the bird-headed Egyptian god Thoth's symbolism pertains to the concept of the Word, which can represent matter in its fully formed state.

Myths of various ancient cultures commonly repeat a metaphor of creation from clay. In Hinduism the goddess Sati-Parvati creates the elephant god Ganesha from clay, then breathes life into him, much like the later fairy-tale story of Geppetto and Pinocchio. In various religious traditions a creator deity is credited with having formed the first human from clay. From a certain perspective the symbolism has parallels in the Dogon tradition, where an animal avatar of humanity is defined as an ant and the Earth is compared to an anthill, which is essentially a mud hill. From the perspective of linguistics, an Egyptian word for "clay," *aqh,* is formed from the same phonetic root as a word that means "to move," "to walk, "to go," and so expression of one concept may have intuitively implied the other.[1]

The notion of ascension lies at the heart of a number of four-stage cosmological metaphors, and the term has significance for each of the three major creational themes. In the context of the microcosm, the term *ascension* refers to the progressive stages by which matter is evoked from waves. However, the idea of personal ascension implies a growth in sensitivity to and understanding of ever-more subtle stages of that same process for an individual. The descending structures of matter lead to Thomson's aether unit, which is characterized as a kind of conceptual gateway between the nonmaterial and material domains. Parallel to that outlook, Buddhist descriptions of ascension in the macrocosm (previously discussed in chapter 9 of my *Point of Origin*) relate to three astronomical effects that direct us to the two Sirius stars. The interaction of these stars mimics virtual particle pairs, whose dynamic, according to Dogon sources, ultimately defines the aether unit as a microcosmic gateway. Cosmological metaphors are framed in various ways to repeat the symbolism of ascending stages of creation. These include four-stage comparisons to the growth of a plant from a seed, and to the life cycle of a goose, which begins as an egg, comparable to symbolism that defines material creation in both the microcosm and macrocosm.

Another of the significant metaphoric themes of the cosmology is given in relation to the concepts of sleeping and wakefulness. In the context of this theme, the nonmaterial domain is compared to a sleeping goddess, while the processes of material creation are described as an awakening. Perhaps the most immediately familiar mythical episode that rests on this theme is the passage that opens the Book of Genesis. The Hebrew god Yah, who is treated in some traditions as the first glow of light from an imminent sunrise, is a glow that hovers over the waters and entices the sleeping goddess to open her eyes and perceive the light. Nonmaterial energy is said to be "of the nature of light," and matter in its wavelike state is compared to water. The sun, which is a source of light, becomes a symbol of the material universe, while the glow of light is commensurate with its energy. From this perspective the Genesis passage repeats the dynamic by which material creation is said to be initiated in the ancient cosmology.

Scientifically speaking, the wavelike behavior of matter is mysteriously transformed into particle-like behavior after an act of perception. Concurrent with this sleep metaphor are various ancient references to eyes. Notions of a left eye or eye of Horus align with the nonmaterial, while those of a right eye or eye of Ra coincide with the material. The eye-shaped vesica piscis, $\bigcirc$, which we identify with Thomson's aether unit, is cast as a gateway between the left and right eyes and so might properly constitute a third eye.

The symbolism of early cosmological architecture often rests on this same metaphor of a sleeping goddess, mimicking a pattern that is epitomized by a famous figurine unearthed in Malta. Its meanings are represented overtly in Dogon cosmologically based architecture and given similar expression in other cultures and locales, most notably in the plan of each of the original houses of the Neolithic Skara Brae village on Orkney. Dogon sources associate elements of the house plan with the goddess's body parts—her head is represented by a round room at one end, her body cavity by a square main room, arms by two rectangular side rooms, and sexual parts by an entryway at the far end. Just after that era in ancient Egypt and elsewhere, it was understood that burial chambers were often patterned after houses, including a First Dynasty burial from Minshat Abu Omar and another from Abydos, reportedly of King Narmer, that replicate what is designated as the "main house" of the Dogon plan. We see in these forms likely symbolic precursors to what Schwaller de Lubicz called the Temple of Man in Luxor, where the structural plan is interpreted as reflecting the anatomy of a human body.

In *Seeking the Primordial,* on a more literal level of interpretation, we compared the four stages of the ascending half-cycle of the Yuga to the stages of human sleep. The cycle of a normal night of slumber culminates with REM sleep, the period of dreaming during which brain activity is high, but the dreamer's ability to physically respond has been suppressed, perhaps to keep them from acting out the events of the dream. These circumstances compare to those described in Samkhya for the fully ascended universe, which, as discussed, is said to have full knowledge but no ability to act. Likewise there is a transfer of energy between the hemispheres of a sleeping brain that compares to the scrolling energy of the two universes. Looked at in this way, references to sleep that may initially seem to have been given as a metaphor take on the appearance of a literal dynamic. This is in keeping with certain other aspects of the cosmology where references we presume to be strictly metaphoric ultimately show themselves to carry more literal significance.

Finally, the dynamic of a metaphor, which provides us with insight into a thing we do not yet perceive through the auspices of another that is more tangibly familiar, again repeats the circumstance of the pair of universes. An underlying goal of the esoteric tradition is to bring humanity to an awareness of a set of effects that, although represented anciently as real, lie largely outside of the range of a modern person's objective observation. The attributes of the nonmaterial domain, like the meanings of the symbolic cosmology itself, are such that while often not provable in the to-five-decimal-points sense, they often can be demonstrated. One purpose of these studies is precisely that—to find ways of demonstrating what an ancient tradition flatly claims to be true.

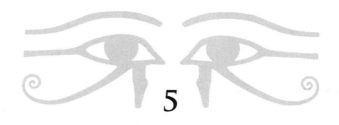

5

THE ALIGNED SHRINE

IN BOTH BUDDHISM AND in the Dogon tradition, an aligned ritual shrine is taken to be both an introductory symbol and the grand symbol of an associated symbolic cosmology. The Buddhist shrine is known as a stupa, while the Dogon shrine—in keeping with a paired civilizing plan whose primary focus is on skills of agriculture—is characterized as a granary. Each tradition emphasizes that the shrine is to be understood as a purely symbolic or conceptual form, not as a practical one—in fact, there is a specific prohibition in Buddhism against putting a stupa shrine to any practical use. Adrian Snodgrass writes in *The Symbolism of the Stupa* that

> To the extent that the building embodies meanings conducive to
> an intellectual vision of the non-duality of the principial Unity and
> manifested multiplicity, it functions as a symbol . . . founded on the
> Indian doctrine that there exists an analogous . . . correspondence
> between the physical and the metaphysical orders of reality.[1]

Researchers have noted that everyday agricultural granaries of the Dogon reflect a very different form from their cosmological granary counterpart. In fact, Dogon specifications for their granary shrine mask an internal contradiction in its dimensions that precludes one from actually being built. Despite outward differences in appearance that we observe between the Dogon and Buddhist shrines in their completed forms, we see fundamental commonality in the base plan and declared symbolism of the Buddhist and Dogon structures. The plans of the two shrines evoke the same set of geometric shapes in the same sequence and assign the same symbolism to each shape. The significant outward structural difference of the two forms is that the Dogon plan culminates in a round base with a squared roof, while the Buddhist shrine classically resolves with a square base and a rounded top. Symbolic aspects of the shrines touch on each of the three creational themes of the cosmology, and so the underlying symbolism can be seen to pertain to creational processes of the microcosm and macrocosm, as well as to those of biological reproduction. We can interpret the structural differences between the two shrines as an outgrowth of the same dynamic of symbolic reversal that we have often noted occurs cross-culturally midway through the twelve-thousand-year half-cycle of the Yuga. We know that circularity is associated symbolically with the nonmaterial domain, while squareness is representative of the material domain. From the perspective of energy that scrolls between those domains, we know that during the first half of the descending phase of the cycle the descending universe would actually be less massive than its twin universe, and so a circle $\bigcirc$, would be its proper icon. Once past the midpoint of the descending phase, that universe would have grown to be more massive than its twin counterpart, and so the figure of a square $\square$ would then become its appropriate symbol. Of course, this outlook carries with it an implication that the Dogon shrine, which rests on a circular base, represents the more archaic form, while the stupa with its squared base would be more recently ancient.

In keeping with its role as a unifying symbol of the cosmology, the esoteric symbolism of each shrine is extensive and has bearing on quite a number of different aspects of the ancient tradition. Accordingly, Adrian Snodgrass's book *The Symbolism of the Stupa,* whose subject turns on those many aspects, runs to more than 450 pages. Similarly, the foundation for the symbolism of the Dogon shrine is developed as part of Griaule & Dieterlen's anthropological study *The Pale Fox,* which also extends to more than 550 pages. At the most essential level of understanding, the symbolism of the Buddhist shrine pertains to physical relationships between the nonmaterial and material domains—and most explicitly to the geometry by which space emerges. In the Buddhist view, the plan of the shrine replicates the process by which space actually comes to be evoked. Such symbolism is telegraphed by the most obvious characteristics of the two shrines. In keeping with the geometry of its base plan, each shrine is aligned to the four cardinal points—a feature that implies a relationship to the symbolic theme of a universal axis. The formation of an axis is one of the effects of the knitting together of electrical and magnetic energy at right angles to each other as electromagnetism—a combined energetic form—emerges. This interlocking effect constitutes an essential "embrace" between nonmaterial and material energies and is what the multitude of ancient "embrace" references likely refer to. The architectural form of each shrine serves to reconcile the geometry of a circle with that of a square (icons of the nonmaterial and material domains, respectively) and so promotes one of the central metaphors for the interrelationship of the universes, which is also characterized as an embrace. Both architectural forms have stairs and so imply the concept of ascension, which constitutes yet another of the key aspects of what's described as spiritually descending and ascending universes.

A Buddhist stupa itself is said to be symbolic of the notion of a primordial mound, set at the conceptual boundary between the nonmaterial and material realms (see fig. 5.1 on p. 58). The method by which this mound is evoked from water recapitulates the process by which mass is evoked from spinning energy that initially takes a wave-like form. The mound is also symbolic of a womb and so simultaneously symbolizes the process by which a womb expands during the stages of biological reproduction. These are mythical themes that are pervasive among many of the cultures that share the cosmological tradition. The same themes are also arguably reflected at the Gobekli Tepe site in Turkey, in its interpretation as a hill with a navel, or in the biologically based symbolism of a clay potbelly. We see the same symbolism reflected in the Egyptian concept of an *akhet* gateway and in the innermost chamber of any Buddhist shrine, which is known as a *garbha*. In Dogon mythology, the mound structure is represented as an anthill into which each of the eight Dogon ancestors are said to enter, headfirst, and from which they are later reborn, as if from a womb. In keeping with that mythical image, ants become symbolic avatars for humanity.

We see evidence of the concept of a primordial mound in the notched passageway of the Queen's Chamber of the Great Pyramid. In its original Buddhist form, the mound is also specifically evoked in rela-

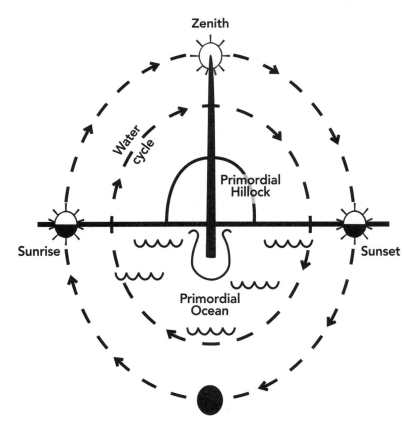

Fig. 5.1. Conception of a Buddhist stupa as the primordial hillock.

tion to an axis. The six arms of the axis, taken together with its center point (which is treated as a seventh direction) correspond to the seven rays of the Dogon po pilu or egg-of-the-world and so also reflect symbolism that relates to the formation of matter.

The image of a mound of earth that was energetically raised up from a body of water in primordial times sets a stage for us to conceive of our material universe in relation to an underworld whose essence is ultimately wavelike. We can easily associate the idea of an unseen underworld with that of an unseen second universe. Likewise, the Buddhist concept of the primordial mound is depicted in relation to a cyclical flow of energies, comparable to the cycle of energy that the Dogon agree persists between the two universes. This outlook naturally associates the material realm with the concept of day and the nonmaterial domain with that of night, so from the very outset there is the strong suggestion that the shrine's plan relates not just to the evocation of space, but also that of time.

As illustrated in fig. 5.2, the plotting of the base plan of the stupa shrine begins with the placement of a stick vertically in a field, around which a circle is drawn, measured with a radius of twice the height of the stick. The initiate then marks the two longest shadows of the day that are cast by the stick (morning and evening) at the points where they intersect the circle. These points of intersection are then adopted as the endpoints of a line that, because the sun rises in the east and sets in the west, automatically takes on an east-west orientation. These same two endpoints are then used again, this time as center points for two new circles of a slightly larger radius than the original circle. These two circles will overlap each other at two new points of intersection, which again become the endpoints of a second line that similarly takes a north-south orientation. With both the sun and the sun glyph as iconic symbols of the material universe, the suggestion is that space emerges in inherent alignment to some broader scheme of primordial compass points.

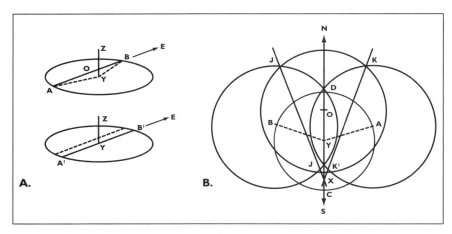

Fig. 5.2. The determination of the orient for planning a ritual stupa
(From Scranton, The Cosmological Origins of Myth and Symbol,
courtesy of Adrian Snodgrass, The Symbolism of the Stupa*)*

The progression of shapes evoked by the alignment plan constitute an introductory course in geometry. These begin with a geometric point and evoke a geometric line as a radius whose purpose is to provide the geometric figure of a circle. The extended plan of the shrine culminates in a geometric square. The measuring out of the two secondary circles effectively constitutes an exercise in bisecting a line. (The invention of geometry is traditionally credited to the Greek mathematician Euclid, at around 300 BCE; however, it is clear that the widespread placement of aligned ritual shrines according to this geometric method far precedes Euclid.) Likewise, the simple act of measuring out the alignment, along with an initiate's use of the cubit as a unit of measure, constitutes an introductory lesson in techniques of construction.

The eminently useful timekeeping functions of the alignment geometry should be immediately obvious to any initiate who actually plots it. The plotting of the first circle with its central gnomon produces an effective sundial and so during the daylight hours gives visibility to the concept time of day in the same way that a modern clock would. (Again, in the traditional view, the sundial was an invention of another ancient Greek known as Theodosius of Bithynia, at around 300 BCE. Clearly the shrine alignment geometry is of much greater antiquity.) The ongoing motions of the east-west alignment line, if plotted daily, give similar visibility to the apparent seasonal motions of the sun. These highlight transitions through the solstices (the points where the line's direction of daily motion reverses) and the equinoxes (points where the plotted line passes through the central stick) and so serve a function that is comparable to that of a calendar.

When measured out in an actual field, the geometry of the plan of the shrine defines directionality in relation to the cartographical concepts of north, south, east, and west. Implementation of the geometry rests on knowledge that we observe the sun rise daily in the east and set in the west and so is reflective of the very concepts of day and night. We could say that even just the initial steps of measuring out the geometric plan for one of these shrines represents an introduction to foundational concepts of geometry, astronomy, geography, measurement and construction, physics, and timekeeping. Meanwhile, the symbolism that is associated with each stage of the alignment process represents a significant introduction to concepts of cosmology. From this very first exercise we could say that ancient instruction in cosmology and civilizing skills went hand in hand with each other in very much the same conceptual way that the dimensions of time and space emerge concurrently. From that perspective, we could argue that this phase of the ancient educational plan itself was symbolic of a cosmological process.

We are told by the Dogon priests that symbolism relating to the macrocosmic theme of the cosmology is evident in the form of the Dogon granary shrine. From this perspective the circular base of the shrine is explicitly symbolic of the sun, while the circle at the center of its square roof represents the moon. The implication is that the body of the granary shrine, which sits between the symbolic sun and moon, represents the Earth. We know that Earth symbolism and Earth-related measures have been long cited as features of the Great Pyramid in Egypt. These go along with other commonalities of form with the Dogon granary that include an entryway that is situated two-thirds of the way up the north face of the shrine. Moreover, the four faces of the granary correspond to four constellations, whose risings and settings serve to regulate the Dogon agricultural cycle by signaling when to plant seeds or harvest a crop, symbolism that again is seen in relation to other world pyramids. Similarly, in cultures such as ancient Egypt, stars were grouped into constellations such that they successively rose above the horizon at a rate of one per hour and so facilitated the telling of time at night, a function that can be seen as a counterpart to the telling of time by sundial during the day.

From another perspective, the moon and sun are symbolic of the feminine nonmaterial and masculine material domains, respectively. It is not uncommon in regions such as the United Kingdom to find ancient stone circles in pairs, one associated with the moon and the other with the sun. Similarly, we take the word "earth" to be symbolic of the concept of mass, and so the term potentially links us with references to the microcosm. We see this most obviously reflected in the inner architecture of the shrine, whose space is segmented into eight chambers—four upper and four lower. The chambers correspond to eight grains of Dogon agriculture and are more broadly symbolic of eight conceptual stages in the formation of the egg-of-the-world. The interior chambers are segmented by two partitions that intersect each other perpendicularly, like an axis. On the ground, at the point of intersection of the partitions, a small cup is placed that holds two grains. The cup is symbolic of the domain of the creator-deity Amma, whose dual nature, symbolized by the grains, is reflected in the combined phonetics of his name (*Am Ma*). Cosmologically speaking, the syllable *am* refers to knowledge (or sexual procreation in the biblical sense), while *ma* refers to an act of perception. These reflect initial acts that catalyze the processes of material and biological creation and so are appropriate to the definition of a creator deity. From this same perspective the shape

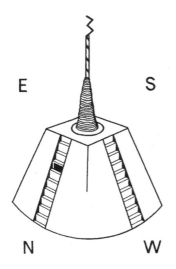

Fig. 5.3. The Dogon Granary: Aligned ritual shrine and central symbol of the Dogon cosmology, comparable to a Buddhist stupa shrine. (Griaule and Dieterlen, The Pale Fox)

of the base of the Dogon granary aligns with their cosmological drawing of the egg-in-a-ball, where the central "egg" is specifically designated as the domain of Amma.

The Dogon granary shrine also enfolds symbolism that is relevant to the cosmological theme of biological creation. We are told that, like various pyramid forms around the world, the shrine can be conceptualized as the expanded womb of a woman who is lying on her back. From this point of view, the two seeds of the cup that is stationed at the pivot point of the internal axis of the shrine can be taken to represent a sperm and an egg. The eight chambers are in keeping with progressive divisions of a fertilized egg, which first evoke two, four, and then eight cells as the egg initially grows. The vertical vector of the axis is understood to symbolize an umbilicus, a biological structure whose purpose is to convey life-giving force in much the same way that scrolling energy does between the two universes, which in the Dogon cosmology are characterized as placentas.

Both the Dogon and Buddhist shrines feature staircases along each of their four faces. In the Buddhist conception, such staircases have an intimate connection to the concept of ascension—a topic that we have mentioned here and discussed in more detail in chapter 9 of *Point of Origin.* The term *ascension* is associated in many cultures with the idea of "climbing up." Ascension in the microcosm relates to the raising up of mass to form particles as a consequence of energy that passes through a kind of gateway between universes. In the macrocosm, ascension pertains to an upward progression that leads to a mythical gateway between the nonmaterial and material universes—seemingly associated with the two Sirius stars. In relation to an individual, the term *ascension* pertains to rising spirituality and is closely linked with the notion of personal enlightenment and the acquisition of esoteric knowledge. So it seems fitting that the Dogon use the four staircases of their shrine conceptually as an organizational tool. It serves as a kind of mnemonic framework for ordering the world in specific hierarchical ways. In the context of discussions of zoology, each face of the shrine is assigned to a specific class of animals (grouped as insects, fish, four-legged animals, and birds, in keeping with the four-stage metaphor), where each step is imagined to represent a particular family or order of related species. The shrine is utilized again in much the same way to conceptualize the various phyla of plants of Dogon culture and so becomes an effective tool for the categorization of important elements of the Dogon world. As previously noted, in relation to agriculture, each face of the granary shrine is assigned to an astronomical body, the risings and settings of which serve to regulate the agricultural cycle and dictate the timing of when to plant or when to harvest crops.

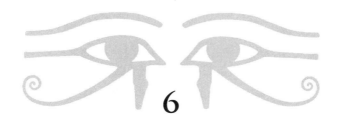

6

ANTHROPOMORPHISM

ANTHROPOMORPHISM IS THE ATTRIBUTION of human characteristics or behaviors to a god, animal, or object. In regard to the ancient creation traditions, anthropomorphism is one of the features that we observe to have emerged over time, in stages. In the root philosophy of Samkhya, whose tenets we also see reflected in the Dogon cosmology, there is arguably no concept either of a deity as we traditionally think of it or of anthropomorphism—only descriptions that give the impression of careful science, stated directly and couched in a set of diverse metaphors. As we move forward historically through the eras of the symbolic cosmology, we see a progression of symbolic forms that seemingly led to the anthropomorphism of deities and ultimately to the notion of gods conceptualized in human form.

The process begins at around 10,000 BCE at Gobekli Tepe with a somewhat unfamiliar set of icons that include stone pillars, enigmatic carved symbols, numerous structurally similar enclosures of unknown purpose, and carved animal images and figures. We can infer from one of the popular translations of the Turkish name Gobekli Tepe, rendered as "Potbelly Hill," that these icons may have included a clay pot that was filled with water to symbolize a womb, which is also known as a potbelly. The carved animal images were likely symbolic of cosmological concepts, in that many of the animals pictured have symbolic significance in later creation traditions, an observation that is supported by comparative linguistics. As an example, we can show that in the

Egyptian hieroglyphic language, ancient names for the animals pictured on the pillars were homonyms for important cosmological terms. Prominent among these archaic stone icons are carved pillars that are traditionally understood to have represented ancestors and so effectively embody one of the earliest known physical expressions of anthropomorphism. In the most obvious case we see a pair of arms and hands clearly depicted, carved in low relief, that reach down to a figurative waistline that is marked by what is widely interpreted as a belt. There is little question that the artistic intention was to partially represent a humanlike figure. Looked at from the perspective of the Dogon cosmology we immediately understand the term *ancestor* to carry cosmological implications. However, based on symbolic elements that survive in a variety of cultures, we can argue that each of the enigmatic carved symbols that coincide with the figure on the same pillar carried cosmological implications. The reasonable inference is that the figure itself was also cosmologically significant.

Because so much of the site is still unexcavated, no one can say with certainty what various other archaic stone icons may be represented at Gobekli Tepe. However, many of the elements that we see there, enshrined in stone, were arguably carried forward as traditional icons of the Sakti Cult and so provide us with conceptual linkage between the two traditions. The matriarchal Sakti Cult survives in India to this day and is thought to have originated as a fertility cult, which means that its connection to the potbelly as a symbolic womb seems sensible. The Sakti Cult was also characterized by hilltop sanctuaries and standing stones set in circles—some researchers speculate with a wooden shrine placed at the center of the circle, comparable to a Buddhist stupa. Naturally, any material object less durable than stone has not survived down through the millennia. Another tradition of the cult was to carve animal images out of natural outcroppings of rock, much as we know was done with the Egyptian Sphinx.

From earliest recorded times in dynastic Egypt and perhaps before, we know that pharaohs, who were thought to have derived their kingly authority from the Egyptian gods, associated themselves with ritual animals, including the falcon, the scorpion, and the serpent, animals that also came to have close associations with concepts of deity. It is through this path of symbolism that animals such as lions later came to be displayed as standards on the shields of warriors. We know that the shrewmouse was also revered in an Egyptian temple at Letopolis, and that other animals were similarly associated with other ancient temples. The word *penu,* which was an Egyptian word for "shrewmouse," also became a generic term for deity in ancient India. The same word appears as a suffix to the names of the two principle goddesses of the Sakti Cult, Dharni Penu and Tana Penu and so can be taken to imply an association with deity even in some contexts where such relationships may not be overt. The suggestion is that by some early era, animals came to symbolize many of the cosmological concepts that we also see associated with concepts of deity. Sometime later, each animal seemingly came to directly represent specific deities, in much the same way that the figure of a vulture is known to have represented the mother goddess Mut in ancient Egypt.

We see a similar dynamic reflected in Egyptian nome signs, icons that were adopted or promoted by regional divisions of ancient Egypt to uniquely identify themselves. Many of these nomes were also known to promote particular regional gods or goddesses. An obvious example of a nome sign is the image of an elephant, which became the highly recognizable standard of the island of Elephantine, located in the south of Egypt at the first cataract of the Nile River. R. A. Schwaller de Lubicz reproduced the image of a boat from Elephantine that displayed an elephant icon. In some cases it seems clear that the choice of icon reflected key products associated with a region, in much the same way that proprietors during the 1700s in the Americas might display an icon associated with their profession in front of or above the door of their shop.

Often in the traditional academic view, animal symbolism is

relegated to the domain of animism, which is often treated as a primitive and unsophisticated spiritual orientation, rather than a considered one. However, from the perspective of comparative cosmological studies, we understand creational themes to have historically preceded any systematic animal symbolism. Concepts of cosmology seem to have provided a foundation for the type of systematic animal and nature symbolism that characterizes many of Jung's archetypes, first and foremost through the association of familiar qualities of given animals with specific cosmological effects. It is in this context that the grasping talons of a bird of prey, the slithering motion of a serpent, the domed shell of a turtle, the reflexive twitching of a rabbit, the spiraling coil of a mouse's tail or an elephant's trunk came to carry cosmological symbolism.

From our perspective, animals that directly symbolized cosmological concepts in archaic times were later, themselves, seemingly deified. Still later, they took on the aspect of avatars or (in the terminology of Egyptian hieroglyphic language) determinatives associated with human or half-human deities. In India, we find a somewhat overwhelming multiplicity of local deities that emerged over time, comparable to local deities in ancient Egypt, but these are largely thought to have been surrogates for a more familiar group of major deities. The animal avatar of any given local deity signals us as to which of the more familiar major gods or goddesses was represented. Cosmologically speaking, it seems likely that the shrewmouse, whose tail we believe corresponded symbolically to a tiny spiral of matter, was later supplanted by the half-human god Ganesha, who had the head of an elephant with a similarly spiraling trunk and whose associated avatar was a mouse. Moreover, our outlook, put forward in the book *Point of Origin*, is that Sakti Cult influences were reflected in ancient Egypt at the island of Elephantine by around 4000 BCE. In keeping with the trend in India, later Egyptian deities were also envisioned to have human bodies and animal heads and so effectively tagged the intended animal symbolism to a god or goddess in another way. The four-stage symbolic metaphor of Dogon cosmology

discussed earlier is in keeping with these practices. The Dogon categories, which are presented in a cosmological context that is remote from
any discussion of deity, also hold true for the symbolism of the animal-
headed Egyptian gods.

Similarly, the clear correspondence of Dogon cosmological concepts
with the mythical roles and symbolism of Egyptian deities is overt and
so implies the influence of an anthropomorphizing effect. The goddess Neith's role in weaving matter is expressly stated and is apparent
in the glyphs used to write her name. The symbolism of the sister goddesses Isis and Nephthys to the stars of Sirius is evident, and we discussed the likely relationship of the interaction of those two stars to a
fundamental dynamic of matter, known as virtual particles, in *Seeking
the Primordial.* Many of the attributes and actions of the Egyptian
god Amen align with those of Amma in the Dogon tradition, who we
can demonstrate reflected cosmological concepts that were seemingly
anthropomorphized as a mythical character.

By the era of the ancient Greeks and Romans, gods had taken on
fully anthropomorphized human form, but often with animal symbolism retained to a much more subtle degree. For example, in Dogon
culture matter is expressly said to be woven by the spider Dada, whose
affectionate name Nana means "mother." Looked at from an archaic
perspective the shape of a spider's body constitutes an "adequate symbol"
for the distinctly spiderlike configuration of an electromagnetic field.
The corresponding Egyptian mother goddess was Net/Neith, whose
express role in Egyptian mythology was to weave matter. We argue that
the glyphs of her name read "weaves 〜〜〜 matter ◠," followed by a
glyph that illustrates angular impulse, a dynamic of energy by which
string theorists say matter is woven, and a symbol that is one of Neith's
icons ⊃⊂. Any spider symbolism that may have originally pertained to
Neith seems to have been lost to ancient Egyptian culture but was evident for goddesses who are traditionally seen as her surrogates in later
cultures, such as the ancient Greek goddess Athena. Numerous other

quasi-anthropomorphized figures populate the mythology of ancient Greece in such familiar forms as centaurs, satyrs, and minotaurs.

Judeo-Christian cosmological symbolism that is directly linked to animals is equivocally evidenced in such characters as the serpent of the Garden of Eden. Modern concepts of deity, as we see represented in Christianity or Judaism, have largely been fully humanized. Even traditional concepts of the Buddha (who in keeping with the archaic view is represented not as a deity but as an enlightened teacher) take human form. We see the flip side of the same coin of anthropomorphism when it is asserted that mankind was made in the image of God—the implication still being that we should think of God in human form. Similarly, the Christian concept of Jesus as essentially a human incarnation of God in its own way repeats Dogon claims for mythical teachers who were said to take action in a material frame but were understood to have originally been nonmaterial.

We often see the phenomenon of deified beings, represented as partly human and partly animal, in many different ancient cultures. Perhaps the earliest of these is seen in the figure of the Urfa Man, from the region of Gobekli Tepe, whose facial features seem distinctly birdlike. There are numerous birdlike figures or representations that have survived from various ancient cultures. Another common blended representation of godlike beings takes the form of partly human and partly fishlike. This is sometimes expressed as a human with the tail of a fish or a figure with the head of a fish over the head of a man, worn almost as a hood. Meanwhile there is an entire school of alternative discussion whose focus is on figures that are described or represented as having attributes of reptiles.

The archaic roots of anthropomorphism are made apparent in a number of ways in Dogon culture. Most immediately we see that the personification of a cosmological concept lends itself to ease of explanation of complex ideas through narrative. Using this approach, concepts that relate to the initiating stages of creation were able to be expressed in terms of the more familiar attributes, actions, and motives of an individual, in this case the deity called Amma. Similarly, aspects of the physics of light could be conveyed through the personal traits and acts of a cosmological character, Ogo. In keeping with the phonetics of the cosmology, root syllables that define various concepts could be adopted as a character's name to reflect the essence of that charater's cosmological role. Those elements taken together provide us with substantial hints as to the subject matter that was being represented. From there, certain cosmological dynamics could then be illustrated through the interrelationship of two or more personified characters. So at root, what we observe with the convention of anthropomorphism is that it opened the door to a set of symbolic techniques whose explanatory potential can be quite powerful.

We often see the same technique applied to Dogon references involving eight quasi-mythical ancestors who are often less individualized than Amma or Ogo. Although the Dogon do apply other names to the ancestors, Griaule most often refers to them using ordinal titles, applying such descriptions as "the eighth ancestor descended before the seventh." Moreover, it becomes perfectly clear that, for the purposes of these cosmological narratives, the ancestors constitute an anthropomorphized device. For example, when Griaule pressed his priestly informant about the storied death of the eighth ancestor, ostensibly killed by the seventh, Ogotemmeli confessed that "no one actually died"—the story line, as framed, was intended to convey a cosmological concept. That said, the eight excavated chambers at the Skara Brae village might uphold a Dogon outlook on ancient instruction; namely, that it was imparted to groupings of eight initiates at a time.

Through Griaule and Dieterlen's final study of the Dogon religion, *The Pale Fox,* we catch the Dogon in the direct act of anthropomorphizing a cosmological construct in their treatment of their egg-of-the-world drawing, or po pilu. The figure is expressed as seven rays of a star of increasing length, which are evoked counterclockwise from a common center point. The rays of this figure are expressly anthropomorphized in two different ways by the Dogon, imagined first as a figure with a head (top two rays), two arms (next two lower rays), two legs, and a single tusk, or alternately with a head and four arms and dancing on a single leg. These stick-figure representations correspond to the two traditional poses of Ganesha and reflect two of Ganesha's iconic attributes—having a single tusk, and the act of dancing. As such, the conceptualizations uphold a relationship we proposed in *Point of Origin* between eight traditional incarnations of Ganesha, which are expressed as progressive stages of creation, and the Dogon concept of the po pilu. Appropriate to those considerations, the word *pil/pille* can mean "elephant" or "tusk" in the comparative terminology of ancient cosmology. Likewise, by literal translation the Dogon concept of the po pilu arguably refers to "the atom of the elephant," a term that is again suggestive of Ganesha. In this quite direct application of anthropomorphism, we get a sense of it as having originated as yet another of the mnemonic devices of the cosmology—a simple method for picturing in our mind's eye the proper configuration of a sophisticated cosmological structure.

7

THE
ROLE OF MYTH

MANY INSIGHTFUL BOOKS HAVE been written about mythology and the role that myths played in ancient times. From perhaps the simplest perspective, a cosmological myth often served as a mnemonic device, an aid to memory whose function it arguably was to make it easier for a person to retain or recall certain information, in much the same way that a fable enshrines a moral lesson for a child. For many modern students of ancient mythology, a common entry point to understanding myth is through Greek and Roman mythology, which for the purposes of our studies represent fairly late forms, somewhat removed from the character of cosmologically based myths that are preserved by societies such as the Dogon. Looked at through the lens of these Greek and Roman tales, mythology can take on the qualities of fairy tales, or of a kind of ancient soap opera whose story lines center largely on the interpersonal dysfunctions of an ancient family of gods and goddesses and so on one level of understanding could also be taken as moral tales. Such stories may have fulfilled some of the same roles for ancient cultures as popular literature does for modern society. For the purposes of these studies, our interest rests with original forms, so our impulse is to trace the conceptual roots of ancient myths. Once again, that impulse leads us to the Dogon and Buddhist myths, whose primary focus is arguably on science, not moral teachings. Discussing the nature of myth, Adrian Snodgrass writes in *The Symbolism of the Stupa,*

> What the architectural symbol is spatially the myth is verbally. The one expresses the supra-physical referent in a geometric or figurative mode, the other in a verbal and narrative mode. As used here the word "myth" is not, as in popular speech, synonymous with "fable," meaning an untrue story. . . . It is a spoken narrative that refers to silence and the inexpressible . . . the narrator of the myth remains silent while speaking. Understood in this sense, myth is "the proper language of metaphysics."[1]

Looking at the Dogon esoteric tradition from a broad perspective, we see two competing motives at work. The first is a studied ethic that carefully shrouds inner esoteric knowledge behind a careful veil of secrecy, one that is only penetrated by the sincere and persistent initiate. For the Dogon, that societal ethic goes so far as to levy penalties against any person (even a priest) who might unthinkingly transgress the proper boundaries of secrecy. The second is a sincere societal desire to attract any interested tribe member to the closely held body of esoteric knowledge. In this second context, public myths (in the form of familiar stories told around a campfire) become a vehicle for introducing cosmological ideas in a very generalized way to uninitiated tribe members. These stories center on broad creational themes, framed in such a way as to introduce important symbolic elements of the cosmology in proper context with one another.

A number of the Dogon stories that are recounted by Marcel Griaule in his book *Conversations with Ogotemmeli* fall into this category of fireside myths. The book itself is organized as a diary of thirty-three days of instruction in Dogon thought that Griaule received from a blind Dogon priest. As a rare outsider introduced to the esoteric tradition, Griaule was granted permission to be instructed as a non-

Dogon initiate and so demonstrated that the tradition is, in practice, open to any sincerely interested person. In his book, Griaule recounts several short myths that were pertinent to his process of initiation. Perhaps the foremost of these compares the sun to a clay pot that has been raised to a high heat, and the planets to pellets of clay that have been tossed outward and scattered around it. To the average listener who may not be well informed about Dogon cosmological ideas, these myths are likely to be received merely as interesting stories, ones that take root in a child's mind and that might later potentially entice the person to think more deeply about the processes of creation. However, to someone who has a broader sense of the Dogon cosmology, the story frames certain elements in proper symbolic or scientific relationship with one another and so gives the distinct appearance of an introduction to esoteric thought. For example, the Dogon description of the sun is roughly in the ballpark with a correct scientific view of the sun's actual character; clay is arguably used in various cultures as a cosmological metaphor for mass; and based on Egyptian glyph symbolism, a clay pot could be a proper cosmological symbol for a microcosmic particle, or for a macrocosmic astronomical body as a conceptual correlate to that particle.

French anthropologists Marcel Griaule and Germaine Dieterlen were careful to explain that there is no grand myth of creation in Dogon culture, nor would we really expect to find one in a society where the symbolic mode of expression is so strongly preferred over a narrative one. Moreover, as we familiarize ourselves with the deeper meanings of the cosmology, we recognize that such complex and nuanced topics do not really lend themselves to expression, all at once, in the form of a single narrative. Contrary to what we may imagine, certain ancient groups such as the Dogon saw the introduction of written language as an undesirable degradation of knowledge, rather than as a technological advance with the potential to further it. The outlook was that written texts represented a potentially dangerous thing. This was, first, because they allowed people who had not truly mastered a subject to represent themselves as experts in it, and next, because a written text, as a physical object, was much less easily secured from profane or prying eyes than a trusted initiate. As compared to the careful and controlled ethic that defines the Dogon oral tradition, any written text that holds significant esoteric knowledge largely becomes a security breach waiting to happen—there is no ultimate guarantee regarding who might ultimately peruse it. Beyond that, knowledge that is conveyed orally from an informant to an initiate isn't subject to the same potential corruptions that might be expected with any written media. Knowledge that is passed from master to future master arguably doesn't suffer from errors of transcription or mistranslation, since it is imparted person-to-person by an authority who thoroughly grasps it, one who is physically present to clarify and explain it. Students who learn directly from a master have an opportunity to actively test the accuracy of their own understanding of any issue that might seem subtle or complex. Moreover, as an effective apprentice to a master, the student also learns firsthand how to properly teach the subject—a skill that no written text can easily impart.

One method that we have seen employed to disguise meaning in ancient written texts seems to have been to substitute homonyms for key words in a written passage. In English, this would be like inserting the word *sea* in place of the word *see*, or *son* in place of *sun*. Looked at phonetically, the import of the passage when spoken aloud remains intact, while for the unknowing translator the message fundamentally changes. Any passage written using this encoding technique could easily be recited aloud by an ancient reader to correctly reproduce the sense of the original passage. However, any attempt to merely translate the passage by someone who is less familiar with the language or unaware of the substitutions would likely render it in ways that ring less than sensibly to the listener's ear. This is an effect that many of us have experienced when we read translations of ancient Egyptian papyri—that what's presented as a careful rendering by a qualified translator ultimately strikes us as nonsensical. Meanwhile, we surely understand the difficulties that translators face when they opt instead to creatively recast an ancient Egyptian passage for the modern ear. More and more often the translator ends up resorting to a poetic interpretation of the text as a way to soften any of the insensibilities that a more literal translation might produce.

When symbology rests on underlying science, the superiority of symbols as a mode of transmission for knowledge becomes even more evident. We see this most clearly in the Buddhist concept of an adequate symbol, which is defined as an icon whose symbolic meaning cannot ultimately be lost because it reflects a shape that is directly observable in nature, such as the highly recognizable shape of an electron orbital ⚛. This category of shapes can be passed along generationally to some future technological audience who will likely recognize them, with no distortion whatsoever of what the shapes represent. As long as the natural shape of the symbol itself is faithfully reproduced, its meaning is ultimately inferable, even when the significance of the symbol may have been enigmatic to any intervening generations—and more so, when comparative ancient outlooks on the symbol uphold the scientific meaning. In other words, there is no "telephone game" effect whatsoever at play with this kind of natural symbol.

We can think of the typical Dogon cosmological myth as an effort to frame a discrete cosmological process in the form of a story line, or to cast it in relation to a set of familiar human actions that serve to illustrate it. In this context, one role of a myth is to convey concepts or ideas that could be problematic to try to frame as discrete symbols. As an example, while the shape created by virtual particle pairs can be illustrated, to properly describe their motion arguably requires some narrative. Most often the technique of relating symbolism through story line is used when there is a desire to convey the subtleties of a process through a working example. This is essentially the same difference of method that we see in modern times when an idea is expressed through the vehicle of a short story, rather than through a painting. However, as we have suggested, the choice to commit an idea to the form of a narrative comes with its inherent downsides. As an example, there is a myth about a Dogon ancestor who steals the fire of his Nummo teachers, with the intention of placing it in the forge of a smithy, which is situated on the roof of a Dogon granary shrine. In his excitement and confusion following the theft, the ancestor makes the rounds of a Dogon granary shrine several times, inexplicably unable to locate a staircase by which to ascend to its roof. The esoteric form of a Dogon granary shrine—the one that is said to house a smithy on its roof—features four such staircases, one situated in the middle of each of its four faces. From an objective standpoint these would be pretty hard to miss. If we endorse the idea that Dogon concepts of material creation rest on the dynamics of angular momentum and angular inertia, then we see that the story line illustrates recognizable stages of those dynamics. These are framed in the context of the ostensible theft of fire, which properly implies the transport of energy, which is precisely what those dynamics are all about.

We could argue that this same story line is preserved in the ancient Greek fable of the Titan Prometheus, who similarly steals fire from the god Zeus. Prometheus, who—like the Dogon ancestors—was associated with the concepts of science and culture, was also credited with gifting humanity with the skill of metalworking, and so the myth also aligns with the Dogon symbolism of the smithy. But if we accept that the two myths seem related, the Greek story line fails to effectively bring forward key elements of the Dogon myth, notably the episodes of the thief running around the circular base of the granary shrine and then ascending—episodes that might lead us to infer a relationship to concepts of spinning energy. Comparison of the two myths illustrates one of the types of uncertainty that a narrative form presents for the intact transmission of cosmological knowledge. Unless each generational storyteller remains scrupulously true to the details of the story line, the essence of the original message can get lost.

The role of myth in a highly stable culture, such as that of the Dogon or ancient Egypt, might not be reflective of its nature for less stable societies or even for stable cultures during potential periods of instability. As an example, various early myths of ancient China seem to encode the definitions of important cosmological terms as episodes or actions that play out as part of a mythical story line. Dressed up in this way, information that we expect to see handed down privately in a society like the Dogon might well have been transmitted in a much more public way in ancient China, disguised as folktales and effectively hidden in plain sight from the uninitiated. We can imagine that if a priestly class in ancient China had foreseen an impending disruption to the generational transmission of esoteric knowledge there, a conscious choice may have been made to disseminate that knowledge in an alternate form. To a certain degree, every cosmologically based myth operates in this same way, in that some creational dynamic is ultimately reflected in the actions or episodes of the myth. As an example, an effect similar to what we observe in ancient China may also be detected in the story of the Exodus from Egypt at the quasi-mythical time of the ten plagues of Egypt. If we allow that the story may rest on an actual historical event, then it obviously also reflects an era of great societal upheaval in Egypt. From that perspective, actions such as the ostensible parting of the Red Sea by Moses with his staff, or of the inundation of a chariot by an energetic flow of water, take on significance that could reasonably be cosmological rather than historical in nature.

8

SYMBOLIC ASPECTS OF ANGULAR MOMENTUM

THROUGHOUT OUR DISCUSSION IN *Seeking the Primordial,* it became increasingly clear that the effects of spinning energy, or angular momentum, lie at the heart of ancient concepts of creation, much as they do in scientific views of matter. Moreover, it seems that the core symbolism of anthropomorphized deities of various traditions also often pertains to the effects of angular momentum. Since this concept seems to be foundational to the symbolic system, it makes sense that we take a moment to clarify how angular momentum and its related effects are reflected by the tradition. In the simplest sense, the processes of material creation begin with two energies that are the components of light, characterized as feminine (nonmaterial) and masculine (material) energies. The engine that drives the processes of material creation seems to be defined by the concept of a dipole, an oscillating effect that arises when two energies of opposite polarity meet, and whose dynamic perpetually brings the positive and negative poles of energy together and apart again, comparable to the action of a beating heart. Science defines several classes of dipole, including electric, magnetic, and gravitational dipoles, whose actions are understood to be rough correlates of one another. Meanwhile, because we understand that the structure of electromagnetism interweaves magnetism and electricity, a reasonable inference is that the feminine energy of the nonmaterial realm is of the nature of magnetism, while the masculine energy of the material realm is akin to electricity. Given that, it seems sensible that the dynamic of a dipole could apply to these as primordial classes of energy. The forms and effects of material creation seemingly emerge as natural by-products of the process by which these two qualitatively different energies come to be entwined in the form of electromagnetism—a process that is also traditionally understood to be catalyzed by the oscillating action of a dipole.

In the ancient tradition, the cosmological context in which these two energies meet is metaphorically compared to the surface (or "face") of water—the boundary at which a body of water interfaces with an expanse of air. So it is not surprising that the dynamics of these energies as they meet might resemble those we also observe on water surfaces. Much as we are familiar with ripples of water that randomly fluctuate in and out, so energetic fluctuations known as virtual particles are thought to persistently emerge and then dissipate at the energetic boundary between the nonmaterial and material domains. Scientifically speaking, virtual particles are a reflection of the interaction of the two energies that crisscross like shoelaces and are part of a dynamic known as angular impulse. As with each underlying dynamic of matter, this one can be seen to play out in parallel form on upward scales of material creation. For example, if we were to graph the positions of two orbiting astronomical bodies such as the moon and the Earth, and consider their motion over time, we would see the same crisscrossing pattern emerge. We have said that any selected segment of that crossing pattern mimics the shape of one of the signature symbols of Neith, the ancient Egyptian mother goddess who is specifically credited with the weaving of matter ⊃⊂. The essence of this same shape also pertains to the Hindu goddess Sati and can be seen represented in Hindu artwork that pictures her. Sati's name rests on the syllables *sa,* which for the Dogon implies the idea of passing a thing hand to hand, and *ti,* which is the Dogon number one, an ordinal number. In various ancient traditions such as Daoism in China, the processes of material creation emerge as the sequence of ordinal numbers, in a metaphorical act of counting. The number one naturally corresponds to archaic concepts of deity. In the later Christian tradition, this idea is represented in the statement that God is One. In the Dogon and Buddhist views, the notion of deity refers to a foundational structure of creation that is not subject to change—effectively the first stage of material creation.

The virtual sea of energy in the cosmos, with its crisscrossing dynamic, creates a kind of scaffold-like structure for the material domain that aligns with the ancient concept of the aether. In the terminology of physicist David W. Thomson III, each almond-shaped segment of this grid constitutes an aether unit $\bigcirc$, a place where energy is ostensibly shaped or configured prior to its transmutation into matter. By Thomson's definition (which aligns with Einstein's contemplations on the concept of the aether), each of these units has a degree of rigidity unto itself, but together they behave collectively like a liquid, much as grains of sand would appear as they pour through an hourglass. The alignment geometry of a Buddhist stupa shrine treats just such a space, known as the vesica piscis, as a place of spacial overlap that emerges between the nonmaterial and material universes. The suggestion is that, because of the relative quickness of time frame that goes along with massless-ness, traditional material laws of physics do not apply within this overlapped space. For the Dogon, this same space corresponds to the egg of the egg-in-a-ball figure $\odot$, the figure that sits at the center of the spinning energy. The Dogon egg is expressly represented as the domain (or "seat") of the Dogon creator deity Amma, whose name was also the affectionate term by which Ganesha referred to his mother, Sati.

Perhaps the first underlying dynamic of material creation results from the tendency of energy to spin whenever two energetic streams of differing quality come together. This spinning action of energy defines the scientific concept of *angular momentum,* and its geometry is correctly represented by the shape of the Dogon egg-in-a-ball symbol. The same shape defines the Egyptian sun glyph and is associated with the

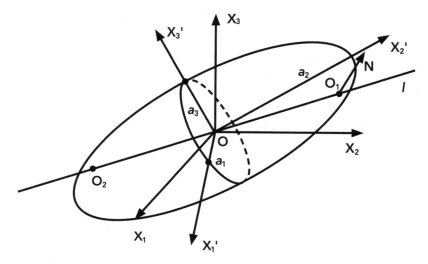

Fig. 8.1. Angular momentum evokes angular inertia: Spinning energy (angular momentum) evokes a fan of vectors (angular inertia) vertical to the plane of spin.

Egyptian sun god Ra—appropriate symbolism given that the sun is the astronomical body around which the other bodies of our solar system also spin. Spinning energy carries with it certain by-product effects, the first of which is to evoke vectors of energy in a plane that sits perpendicular to the spinning energy. This effect is known as angular inertia.

The evoked vectors of energy are conceptual correlates to the seven rays of a star of increasing length that characterize the Dogon *po pilu*, or egg-of-the-world. It is this extended construct of spinning energy, with its evoked vectors, that we take to be symbolized by the Hindu elephant god Ganesha. The oscillation of the spinning energy is traditionally understood to be that of a dipole, and its action is what causes magnetic and electric energies to be interwoven in a perpendicular relationship to one another. That perpendicular construction defines a conceptual axis and so reflects one of the core concepts of the ancient cosmology. The combined form of the two energies is known as electromagnetism and is traditionally seen as one of the essential building blocks of our material universe.

As discussed in chapter 15 of *Point of Origin* ("The Cosmological

Role of Ganesha"), features of this basic dynamic of energy play out as attributes, icons, and incarnations of Ganesha. Like all good cosmological terms, a phonetic reading of the name of Ganesha itself, taken in archaic form as Ga Nu Sa, implies the concept, "temporality (*ga*) and waves (*nu*) attain knowledge of each other (*sa*)," definitions that well describe the net effect produced by the spinning energy. As we should expect, a number of the key attributes traditionally assigned to Ganesha, as expressed through various myths, also reflect the dynamics of angular momentum. Most obviously, Ganesha spins, just as angular momentum does. One of the essential roles of Ganesha is to impose and remove obstacles. The notion of an obstacle that is first encountered but then overcome defines the scientific concept of *resistance*. The indication is that the progressive lengthening of the vectored rays of angular inertia described by the Dogon is actually a by-product of resistance that is evoked as the energy spins. Another scientific concept, known as the *polar moment of inertia,* is an indication of rigidity, and the *mass moment of inertia* is the resistance of a massive object that accrues when we rotate it. Such resistance is described quantitatively by the term *polar moment of resistance.* When the pertinent quantities are graphed, they produce a figure that outwardly resembles the shape of the head of an

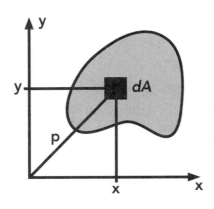

Fig. 8.2a. Graph depicting the moment of inertia.

Fig. 8.2b. An elephant head resembling the graph figure.

elephant in profile, which is also one of the iconic attributes of Ganesha.

The figure of an arrow or knife, which Ganesha often holds, is representative of the vectored energy that characterizes angular inertia. The scepter he often holds replicates a figure astrophysicists use to represent increasing resistance that accrues with distance from the plane of the spinning energy. The noose that Ganesha is sometimes pictured holding reflects one of the shapes that resistance causes energy to take, seen more obviously in the orbital configurations of an electron. The nested petals of the lotus flower reflect the shape of nested cones of energy that characterize the vector model of angular momentum. The Dogon spiral of matter, which may correspond to Ganesha's often spiraling trunk, is an alternate way to conceive of the vectored rays of angular impulse, perhaps as seen from a higher dimensional view. In some contexts Ganesha is shown holding a radish. The symbols of an Egyptian word *kau* for "radish" suggests that it represents the center point of emergence of the spiral of matter. The elephant god's *vahana,* or mode of transport, is represented as a rat, symbolism that likely pertains to the animal's quickness and stealth. These attributes may reflect the initially quick time frame that underlies and evokes the vectors of angular inertia. The rays of angular inertia are the product of the resonance and resistance that spinning energy evokes. Each of the seven rays corresponds to an instance of resonance, and so there would be seven resonant structures metaphorically subsidiary to Ganesha. We take these to be represented in Hinduism by the concept of *ganas,* the "flock" or "multitude" over which Ganesha (as Ganapati) was understood to be the leader. The signature shape of this component of resonance is reflected in one traditional capital of *amalaka* pillars, a common iconographic element in Buddhist art and architecture.

In support of this outlook, Adrian Snodgrass tells us that "in some [instances these] four thin corner posts are replaced by [draped figures] of ganas."[1] One aspect of the ganas is that they attended ritual sacrifices known as *vajna*. However, another meaning of the term *vajna* (which Ghandi called the "true vajna") was spinning—precisely the action of the seven associated resonances. The Dogon term *gana* refers to the concept of space and can mean "to cleave," which means to divide. The Egyptian root ga means "to overturn" and so also reflects the concept of spinning. The related term *gah* means "to reach out," "to stretch out," "to extend" and so also reflects the concept of the expansion of space.

The second underlying principle of physics that applies to our example, one that also aligns with Einstein's theory of relativity, is the understanding that when we accelerate energy, we impart relativistic mass to it. According to Einstein, as an energetic body becomes more massive its relative time frame slows. However, Einstein also emphasized that the speed of light itself remains constant without regard to changes in relative time frame, which of course implies that light takes its origin outside of the context of material time frames. Since speed is calculated simply as distance divided by time, then from the perspective of an outside observer, for the speed of this light to remain mathematically unchanged, the progressively slowing time frame of spinning energy must correspondingly evoke vectored rays of ever-increasing length—precisely as the Dogon represent them. What the Samkhya philosophy describes as a dimensional dynamic that fosters seven sets of paired universes might reasonably configure in ways comparable to those as seven rays, much as sound expresses itself as seven progressive musical tones and light as progressive wavelengths of light, which we perceive as the seven colors of the spectrum. The suggestion is that any energetic impulse that crosses the boundary from the nonmaterial to the material universe may undergo a kind of transformation in keeping with this one-to-seven dynamic. The suggestion is that the movement from Unity to Multiplicity may constitute an inherent transmutational effect. This is the same rationale that might explain the clusters of meanings that consistently attend concepts of cosmology in many ancient languages.

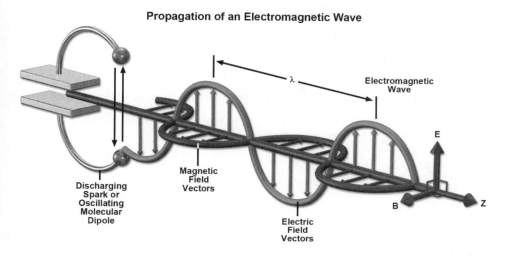

Fig. 8.3. The electric field in an electromagnetic wave vibrates with its vectorial force growing stronger and then weaker, pointing in one direction, and then in the other direction, alternating in a sinusoidal pattern. At the same frequency, the magnetic field oscillates perpendicular to the electric field. The electric and magnetic vectors, reflecting the amplitude and the vibration directions of the two waves, are oriented perpendicular to each other and to the direction of wave propagation. (Image by MolecularExpressions at Florida State University; used with permission)

The implication is that, much as the Dogon priests assert, the root concepts of creation reside ultimately on the nonmaterial side of the equation.

If, as the Dogon conceptualize things symbolically, the seven vectored rays of the po pilu constitute the Word of matter, then we might imagine that time frame, which we interpret as the source of resistance of energy, similarly constitutes its voice. The directional arrow of time is the scepter that is translated and conveyed from the non-material domain to the material one. If we imagine that the metaphoric voice of matter expressed itself most loudly in the archaic eras of the descending cycle of the Yuga, then it becomes understandable that a lion—the member of the animal kingdom endowed with what is argu-ably the loudest voice (or roar)—came to be its symbol. In keeping with this outlook an Egyptian term for the Sphinx at Giza was Hu,[2] while a term for a scepter was also *hu*.[3] Symbolically, this word for scepter reads as "dimensions" ⚱ and "grow," 🐦, followed by the image of a downward-pointing arrow, symbolic of the descending half-cycle of shifting time frames. This shift is traditionally understood as the collapse of the wave function that distinguishes matter in its wavelike state from our familiar material domain. It is this same set of concepts that defines the symbolism of Siva (pronounced "Shiva"), who was the consort of the Hindu goddess Sati (or Parvati). These are at the root of the effects that effectively differentiate particles from waves, and whose effect compares metaphorically to the action of a sieve.

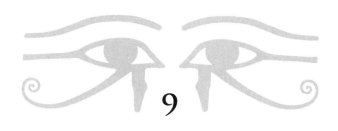

9

SYMBOLISM OF
TIME AND SPACE

ALTHOUGH IT MAY NOT be immediately obvious to the student of ancient cosmology, symbolism that has bearing on the concepts of time and space is foundational to the ancient cosmology. We first see this with the same introductory exercise for initiates of Buddhism whose purpose is to align a ritual shrine to the four cardinal points. The goal, cosmologically speaking, is for the initiate to master the geometric method by which Adrian Snodgrass says the "sacred space" of the shrine is differentiated from the "profane space" of a disordered field. We have said that this is a process that defines the base plan of an aligned stupa shrine, one that is also expressly associated with the idea of the evocation of space in the material universe (see fig. 5.2 on p. 60). Buddhist sources flatly tell us that the method recapitulates the stages by which space actually comes to be evoked. Snodgrass writes that Buddhism "contains many references, both implicit and explicit, to the concept of the directions of space deploying from a centre," which is precisely how the alignment geometry expresses itself.[1] In *Seeking the Primordial,* we inferred that these geometric figures and relationships, which lie at the heart of the concept of *sacred geometry,* actually constitute far more than a mere symbolic representation. In truth, many important clues to the essential nature of underlying cosmic relationships and creational processes seem to rest with this geometry.

The symbolism of this geometry is significant enough to warrant a recap of how it pertains to the concepts of space and time. We have said that the alignment geometry of a stupa shrine also outwardly facilitates two related timekeeping functions that can be seen as pre-requisite to success in agriculture. The first phase of stupa alignment involves the placement of a stick vertically in a field and then the tracing of a circle around it. Here, the stick's functional purpose is that of a gnomon, simply to cast shadows in the sun. We know that this is the essential configuration of a sundial, a tool that allows one to keep time during the daylight hours. A second by-product of the plan of the alignment geometry, which is evoked during the next stage of the initiate's process, is that of a functional calendar and makes the motions of the sun apparent during the course of a year and so gives visibility to the dates of the solstices and equinoxes. The calendric function is accomplished by marking the stick's two longest shadows of the day, morning and evening, at the points where they intersect the drawn circle. Since the rotation of the Earth makes the sun seem to rise in the east and set in the west, these two points automatically define a line that will take on an east-west orientation. If an initiate were to mark a similar line each day for a year, then over time the line's placement could be seen to oscillate northward until the date of the next solstice, turn southward again and pass through the central stick on the dates of the next equinox, then continue farther southward until the date of the next solstice. Working simply from this one exercise, an initiate could easily work out the correct length of a solar year and forecast changes of season simply by counting the days that elapse between solar events that are now made visible. As the next phase of the geometric plan, the initiate adopts the endpoints of this east-west plotted line as the center points of two additional circles, drawn with a radius somewhat greater than the first. The larger radius causes the circumferences of these circles to overlap each other at two new points of intersection. These are again taken as the endpoints of a second line, one that will sit perpendicularly to the first line, and so will automatically take on north-south directionality.

The two timekeeping functions evoked by this introductory geometry, those of a clock and a calendar, each pertain specifically to the practice of agriculture and so have direct application to the associated civilizing plan, whose primary purpose is said by the Dogon priests to have been to foster skills necessary for agriculture. Geometry that, from one perspective, reflects how space actually emerges was framed to evoke two practical tools of agriculture. In this way, cosmological symbolism came to be intimately entwined with instructed civilizing skills. This kind of intimate interrelationship between cosmological concepts and practical civilizing skills can be seen as a signature characteristic of the plan of the ancient cosmology.

Popular science tells us that space and time (Einstein's space-time) are inextricably linked to each other, but from an everyday perspective it can be difficult to conceptualize the idea as much more than an abstraction. Even when first working with Dogon descriptions of how matter is said to be evoked in a comparative context, our impulse was to sort out at which stage in the creational process the concepts of space and time make their first appearance—to essentially wonder which effect the Dogon represent as having preceded the other. However, the true situation appears to be a more subtle one than that. The emergence of space and time gives the appearance of being more like that of the leavening action that causes bread to rise, in that the two codependent effects seem to be inextricably interlinked. Bread traditionally rises and expands over a period of time, hand in hand with a leavening effect that generates and traps gas within the dough. Whereas we might imagine nonmateriality to be a condition that precedes the concept of time, we instead come to see the nonmaterial oscillation of time as a preexisting dimension, where time frame runs ultra-quickly compared to the one we experience. Ultimately we see that it is the influence of this ultra-quick time frame that produces the otherwise confusing effects that we observe both in the quantum world and with entangled particles. The quality of universal entanglement that should accompany this ultra-quick oscillation would give an outward appearance of Unity to all things. Localities would seem so immediately interlinked that traditional notions of space and time would no longer have any effective meaning. So to imagine, as we did, that the nonmaterial domain somehow lacks a dimension of time would seem to be a false perspective.

We arrive at the likely truth of the situation through Einstein's concept of relativity, which rests on a constant speed of light—a notion that should go against our commonsense perception. Imagine that you are driving on a highway alongside a train that is moving somewhat faster than you are. Increase your speed, and your car will begin to catch up with the train. According to Einstein, that scenario of accelerating to catch up does not hold true for the speed of light. No matter how much faster you might drive your car, you would never even begin to catch up with the speed of light as measured relative to your motion. Since speed is calculated as the distance a thing travels divided by the time spent traveling (as miles or kilometers per hour), the only way, mathematically speaking, that the speed of light can remain constant while you actively speed up would be if hours somehow grew longer—your time frame would have to slow down in proportion to how much you accelerate. Einstein tells us that the effects of accelerating a material body are indistinguishable from those of its growing more massive. His theory of relativity also tells us that time is a relative concept—that as the mass of the body increases, its time frame correspondingly slows down. A more familiar metaphor by which to picture this would be to consider the effect that increasing depth of water has on water pressure, and how that, in turn, affects a buoyant object. Just as the force of the pressure of water automatically increases on an object the deeper we go, so the time frame we experience automatically grows slower the closer we move to a body with more mass. A buoyant object gravitates to regions of lower pressure in the same way that objects of mass gravitate toward regions of slower time frame. Now if we think of this relationship between massiveness and time frame as a continuum, then as the scale shifts toward the side of greater mass, time frame progressively slows down; if we shift the scale in the opposite direction toward the side with less mass, time frame must correspondingly quicken. This trend is both a continuous and a linear one, in that there is no clear opportunity along the way for the dynamic of the relationship to change. This implies that any cosmological structure or domain that is virtually massless must also experience a time frame that is ultra-quick compared to ours. Such an outlook implies that the processes of material creation we observe do not rest on the emergence of time as a dimension so much as they do on a progressive slowing of time frame that causes the effect of time to be more pertinent. In that context, it makes sense to ask ourselves what the concept of space must be like in a virtually massless domain where time frame has dramatically quickened.

Even though Einstein tells us that time frame slows as we accelerate, we experience no obvious sense of change in time frame in our everyday lives, at least not that we readily recognize. In truth, from our perspective the very tangible effects we feel of momentum and centrifugal force might well be a consequence of slight differences in time frame, but we may simply have never conceptualized them as such. So the suggestion is that the effects of relative quickness of time frame may not be readily apparent to those who are in it—it may be that these only truly become evident from the standpoint of an outside observer. As an illustration of this idea, imagine that you want to transport yourself 150 miles southward from Albany, New York, to Manhattan. There are any number of ways to potentially do that, and the travel time required can differ greatly based on the method of travel you choose. Among other possible methods you could go by skateboard, by bicycle, by bus, by car, by train, by conventional airplane, or by jet airplane. Among these modes of travel, your travel time might vary from several days to under an hour. Meanwhile, with today's technology, simple communication with someone at that distance by phone or over the internet is effectively instantaneous. In any case, the objective distance between the two geographic points remains constant regardless of which method of travel or interaction we choose—the true distance is always measured at approximately 150 miles. However, the effective distance—especially from the perception of an outside observer—shrinks dramatically as we employ progressively quicker modes of travel or interaction. The actual distance of 150 miles doesn't physically shrink regardless of our choice, but the pertinence of that distance on the interaction does. If all modes of interaction became instantaneous, then the physical fact of the distance would become moot. Looked at in this way, it is possible that on the scale of universes, distance (like the speed of light) remains a constant, but its pertinence diminishes as time frame quickens.

Speaking both mathematically and pragmatically, if it takes us no time to move from point A to point B, then whatever distance might objectively separate the two points has no ultimate bearing. The net consequences for us in a virtually massless universe would be the same as if there were no distance at all between the points. In other words, an ultra-quick time frame would create a circumstance similar to what we observe with quantum entanglement, where two particles (such as two electrons) can be induced to behave as if they were effectively one, despite however great a span of distance might seem to separate them. Looked at in this way, the processes of material creation rest on a fundamental difference in time frame that, based on Einstein's view of relativity, must distinguish a virtually nonmaterial domain from a more material one. The scientifically testable circumstance of quantum entanglement actually demonstrates the likely truth of the outlook, since the otherwise confounding effects of entanglement can actually make sense within the context of an ultra-quick time frame. When we entangle particles we change some factor of their relationship, one that introduces a situation where traditional laws of time and space seem to temporarily break down. The symbolism of ancient cosmology implies that this introduced factor should be understood as an ultimate quickness of time frame compared to ours—a circumstance that we believe also characterizes the overlapping spaces of the aether units.

To that point, perhaps the most pertinent symbolic element of the stupa alignment geometry is the almond-shaped region of overlap that we said is evoked between the secondary circles that are drawn by the initiate. This is the anciently revered figure of the vesica piscis (literally, the "bladder of the fish"). From one perspective it is also what is represented by the Egyptian mouth glyph ⬭. If we endorse the Buddhist perspective that the stupa alignment geometry illustrates how space actually emerges, then by implication this figure would replicate a fundamental overlap that occurs between the nonmaterial domain and our material universe. As we understand things, this space becomes the focus of root concepts of cosmology in several ways. First and foremost, given the extended metaphor of an embrace between universes that characterizes many symbolic features of the cosmology, it is this geometry that physically constitutes that embrace. Moreover, based on the four-stage agriculturally based metaphoric progression of the Dogon, the shape of the vesica piscis corresponds to a variety of millet seed called the *yu*, which the Dogon say represents "the smallest seed." As a metaphor, this seed represents a starting point for concepts of material creation. It is from this structure that the material effects of temporality (past, present, and future) are evoked. These temporal concepts are expressed by the Dogon syllable *ga*, and so taken together with the yu seed seem more than suggestive of the concept of the Yuga. From another perspective the vesica piscis is the space that comes to be enclosed by Amma's clavicles and so becomes the conceptual vessel of what the hidden god "grasps," "holds firm," or "establishes."

As we mentioned in *Seeking the Primordial,* the Dogon outlook on foundational processes of matter corresponds to a little-known theory of astrophysics called the Aether Physics Model, which was formulated by David W. Thomson III. Thomson calls his correlate to this almond space the aether unit, a place situated at each point of space-time where primordial energy comes to be wrapped and configured, prerequisite to the processes of material creation. Thomson infers that time as it persists in a nonmaterial context represents an oscillation. This outlook is upheld in Buddhism where one dynamic of time is symbolized by the clapper of a bell that swings back and forth as if to illustrate the concept of an oscillation. Similarly (and perhaps contrary to expectation) the Egyptian hieroglyphic language as represented by Budge defines two distinct glyphs to represent the concept of time. The first takes the form of a circle that encloses two parallel lines and so conjures both the plotted-line geometry of the stupa plan and the notion of an oscillation Ⓘ. The second resembles one of both Thomson's and the Dogon's initiating configurations of energy, which the Dogon priests refer to as a guide sign. Thomson identifies the figure as a loxodrome ʃ. Thomson considers that the nonmaterial universe sees only the forward half of the oscillating rotation of time and so infers from that the presence of an extra dimension of time for the nonmaterial universe. From our perspective, the rotating oscillation of nonmaterial time has the character of a dipole and so evokes a directional vector of time that corresponds conceptually both to a dipole moment and to the dynamics of angular inertia. From that perspective, it would only be the nature of how the dimension of time is expressed (rotationally versus directionally), along with its relative quickness, that differs between nonmaterial and material frames. From this perspective, no additional dimension of time need be postulated, only a dimensional quality that comes to be translated.

Looked at from a more essential viewpoint, the geometry of the stupa alignment plan reconciles the dynamics of these two modes of time that are respectively associated with the nonmaterial and material universes. The sundial aspect of the geometry reflects the character of time from our material perspective, while the motions of the oscillating seasonal line of the plan mimic the dynamic of time as we believe it to be experienced from a nonmaterial perspective. Just as the stupa plan can be seen to reconcile the geometry of the two universes and so effectively depicts how space comes to be evoked, so the geometry also reconciles the two distinctive dynamics of time. Of course, this is as it should be, since we understand the material dynamics of space and time to arise concurrently through the same progressive slowing of time frame that is induced by the evocation of mass.

The intuitive cosmological symbol for space is a geometric square □. Because our material universe exists in four dimensions, the four-sided figure of a square is—reasonably enough—its symbol. Quickness of time frame causes the nonmaterial universe to take on the outward appearance of Unity, which should be sensibly represented by a single-sided figure, and so a geometric circle becomes its icon ○. The processes of material creation arise from what, based on the alignment geometry of the stupa, is a reconciliation of nonmaterial and material aspects, so from one perspective its iconography rests metaphorically on that notion. The figure of a hemisphere ◠, which we take to represent mass or matter, reconciles a circular shape of the nonmaterial with the square of materiality. The same Egyptian glyph is interpreted in some contexts to represent a loaf of bread and so implies the very same leavening process that we compared to the emergence of space and time. The ubiquitous cosmological symbols of a compass and a carpenter's square constitute tools for producing those figures. Similarly, we have noted that in archaic times the figure of a hemisphere or dome came to signify a sanctuary or temple, a place where nonmaterial and material energies were understood to come together.

The concept of time, as it is understood in Buddhism, is alternately represented as a wheel (sometimes referred to as the Dharma Wheel) and so is configured from one perspective as an oscillation, just as Thomson characterizes it. On another level of interpretation, the spokes of the wheel correspond conceptually to the shadows cast by a sundial and so also pertain to the directional arrow of time as we experience it materially. The wheel itself is associated with a chariot and so is linked cosmologically with the macrocosmic spiral of Barnard's Loop, a structure that the Dogon refer to as the Chariot of Orion. Such an association is also in keeping with the idea that spiraling energy becomes the conveyer of mass and time frame between the two universes. In Buddhist thought, the number of spokes depicted in any given representation of a Dharma Wheel has bearing on its intended symbolism. According to Snodgrass, a four-spoked wheel "is a model of the solar-structured world"[2] and is in keeping with the four-dimensionality of our material universe. The six-spoked wheel represents the sun and is in keeping with the six directional branches of the axis of the universe (the center point often taken as a seventh direction). An eight-spoked wheel signifies the eight directions of space and is also taken as a wheel of renewal and regeneration. This symbolism coincides with the eight stages of the Yuga Cycle (four descending and four ascending). A twelve-spoked wheel pertains to the months of the calendar year and divisions of the zodiac. These definitions span the range of symbolism that we associate with the concept of time based on archaic references.

In keeping with the character of a sundial, the image of a spoke, scepter, or arrow symbolically implies the directional vector of time that we experience materially. As we have suggested, the vector of time is another of the tangible effects of the spinning energy of angular momentum. The symbolic icon of an arrow is also typically associated with Mother Goddesses such as Tana Penu of the Sakti Cult and later surrogates of the Egyptian goddess Neith, such as the Greek goddess Athena. However, mythical plotlines that depict the use of arrows by a Mother Goddess in India often also focus on influences of deities like Ganesha. One source says of Ganesha:

Ganesha is the first sound, OM, in which all hymns were born. When Shakti (Energy / Matter) and Shiva (Being / Consciousness) meet, both Sound (Ganesha) and Light (Skanda) were born. He represents the perfect equilibrium between force and kindness and between power and beauty. He also symbolizes the discriminative capacities which provide the ability to perceive distinctions between truth and illusion, the real and the unreal.[3]

Ganesha himself is understood to be the placer and the remover of obstacles and is associated with the intriguing concept of a *productive obstacle*—the notion of deflection for a productive purpose. In Dogon culture the action of the spiral of the po pilu (angular inertia) is compared to that of a sieve, which is utilized by the Dogon as a productive obstacle to separate beans from sand, in which they have been placed as a way to preserve them. In one Hindu myth, Ganesha is portrayed as repeatedly deflecting his mother Parvati's arrows. One possible implication of the myth is that the influence of certain of the seven vectored rays of the po pilu may be deflected from direct material expression by the dynamics of angular inertia.

Once Parvati was out hunting deer with her son. She took careful aim, but just as she was about to release the arrow, Ganesha knocked against her arm. This happened repeatedly. Parvati turned to him in exasperation. He smiled, and reminded her, "Mother, you have forgotten to invoke your son. Before you take aim you must utter, 'Om shri Ganesha namaha.' Then and only then will your arrow hit its target."[4]

The word *om* refers to energy or sound, and the word *shri* comes from a root that means "seed." Based on the meanings of phonetic roots of the cosmology, the word *namaha* implies the notion of "the feminine, perceived and ascended." Taken together, the references imply a dynamic in which only certain energetic impulses associated with the arrow of time actually pass through to the material domain. The suggestion is that some may be deflected by obstacles associated with the effects of angular inertia. This outlook could go hand in hand with our perspective on the half-cycle of the Yuga, where, due to differences in time frame at the extremes of the half-cycle, passage between the less material and more material universes seems to be indicated only during the middlemost eras, most particularly at what we might interpret as the equinoxes of the Great Year.

As stated, we understand from Dogon references that Amma resides in the central egg of the egg-in-a-ball figure. This is the midpoint of angular momentum, around which energy spins. If we take an arrow (the vector of angular inertia) to symbolize the concept of our directional vector of time, and Ganesha to represent the concept of angular inertia, then the suggestion is that angular inertia imposes and removes deflective obstacles to that directional vector. This makes sense scientifically, since we know that when light passes the boundary between two different mediums (say, air and water) its direction can be bent—except when the path of light is perpendicular to the medium it enters. We know that the Egyptian dung beetle Kheper, whose name also was a term for the equinox, represented the concept of nonexistence coming into existence. The equinoxes marked the center point of the alignment of the stupa shrine geometry and so correspond to the centermost vectored ray of angular inertia—the one situated perpendicular to the center point of the spinning energy. We also know that the seven colors of a rainbow represent seven vibrational deflections of white light, and that white is the color associated cosmologically with the nonmaterial. Green, which in certain symbolic contexts associates with the material universe, is the fourth of seven colors and so sits similarly midway through the sequence of a rainbow's frequencies. Dogon and Egyptian words for the color green rest on a phonetic root that can mean "to come" or "to travel from one place to another." The Dogon term *veru* combines two phonetic roots that can imply, cosmologically speaking, the concept of *nothingness dispersing.*[5] Similarly, our material universe is defined by the Dogon priests as the fourth of seven material universes, which also positions it midway in the grouping. There is a consistent pattern reflected in this symbolism that may have pertinence to how possibilities, if rooted in the nonmaterial, may unfold as material events. The suggestion is that angular inertia plays the role of a sievelike obstacle to filter out certain possibilities that might come to be realized in a material frame. If so, then alignments between the dynamics of the two universal time frames might affect which possibilities come to be realized materially.

In keeping with Dogon phonetics, many of these same concepts pertain to the cosmological notion of a Yuga Cycle. The Hindu and Buddhist concept of the Yuga Cycle aligns with the broadest cycle of time that humanity directly observes, known as the precession of the equinoxes. From an earthly perspective, this is the cycle in which the constellations of stars seem to slowly rotate around us over a period of approximately twenty-five thousand years, like some grand clockwork. The cosmological outlook is that humanity's ability to perceive the nonmaterial domain fluctuates during this cycle, decreasing (descending) during a half-cycle of more than twelve thousand years and then increasing (ascending) for another twelve thousand years. Once again, these are effects that, based on the Dogon and Samkhya view of cosmology, accrue from the progressive shifting of time frames. As primordial energy scrolls from one universe to the other to the limits of its extent, mass effectively scrolls with it. Mass decreases in the less material domain, while its relative time frame quickens. Similarly, the time frame of the increasingly material domain progressively slows. We infer that the decreased ability on the material side to perceive things nonmaterial, as described within the Yuga Cycle, must accrue from the ever-growing disparity of time frames between the two universes. The physical effect might again be similar to that of a suspect in a police interrogation room who is unable to see the people who may peer in at them.

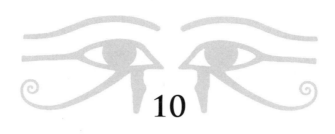

10

MYTHOLOGY
OF LIGHT

AMONG ANCIENT CULTURES, LIGHT was a quality that was almost uniformly associated with concepts of the nonmaterial, spirituality, and deity. In Buddhism, white light was an icon of Unity, in that it was seen to subsume within itself each of the other colors of the visible spectrum of light. White is so quintessentially treated as symbolic of the nonmaterial feminine that in 1948 Robert Graves enshrined the idea in his book-length essay *The White Goddess: A Historical Grammar of Poetic Myth*. Similarly, in *The Symbolism of the Stupa* Adrian Snodgrass reproduces a diagram of a Buddhist color wheel that features white at its center point as a symbol of the nonmaterial and frames the primary and secondary colors of the rainbow around the wheel in proper relationship to each other (orange between red and yellow, green between yellow and blue). This diagram inscribes within a circle a figure that takes the traditional form of a Jewish Star of David, comprised of two equilateral triangles—one pointing upward and the other pointing downward.[1] Much as we know that the broader spectrum of light includes frequencies that the unaided human eye does not readily perceive, so the nonmaterial realm is treated as an energetic domain that falls largely outside of human sense perception.

The feminine and masculine energies of the nonmaterial and

material universes, whose interactions provide the foundation for angular momentum and its resulting structural forms, also produce at least one other essential cosmological by-product; namely, light itself. The extent to which light is emphasized as a mythical concept is perhaps most obviously seen in the Book of Genesis of the Old Testament, where the very first act of the Hebrew God Yah (or Yahweh) is to proclaim the existence of light, or perhaps catalyze its first perception. By one familiar cosmological metaphor the nonmaterial universe is characterized as a sleeping goddess and the processes of material creation as an awakening. Adrian Snodgrass also informs us of a Buddhist perspective on the relationship of light to concepts of material creation that is consonant both with the biblical one and with the Dogon conception of an egg-of-the-world. He writes:

> The Golden Womb or Egg is the progenitive point where the ray of light or radius of Breath strikes the plane of the waters. . . . When the Breath enters or the Light shines upon the waters the possibilities they contain are made manifest.[2]

Like the Dogon, Snodgrass understands the structures of material creation to arise from the joining of two principles of essence and substance, which themselves represent a "diremption" (division into two) of an even more essential state of nonbeing. He writes: "The analogy for this productive conjunction of principles, and one that is to be found in many traditions, is given by the image of light reflecting from water."[3] Similar meanings are apparent in the Dogon term for light, *yalu*, which rests on the same phonetic root *ya* and also implies a meaning of "the first glow of light at dawn."[4] Likewise the Dogon term *ya* implies the concept of femininity,[5] which we know is a defining attribute of the nonmaterial. For the Dogon the term also reflects the concept of *an acquired state*, such as that of mass, which is fostered through the joined energies of electromagnetism as it spins.

Commensurate with all of this, the Dogon word *yalu* also refers to the notion of place[6] and so reflects both the cosmological concept of space and the geometric notion of a point that go hand in hand with the emergence of light.

Contrasts involving light underlie some of the most basic oppositions of the symbolic cosmology, and these can be understood on one level of interpretation to reflect the opposing characters of the nonmaterial and material universes. By such metaphors, mass is properly framed in conceptual opposition to energy. First among these would be the archetypal separation of light and darkness, or day and night. We see this same fundamental pairing presented by early sister goddesses, such as Isis and Nephthys in ancient Egypt, whose symbolism is to the light Sirius star and arguably to its dark, massive companion. A similar symbolic pairing is seen in India with the sister goddesses Dharni Penu and Tana Penu of the Sakti Cult. The word *dharni* means "luminous" and so evokes the concept of light, while the phonetics of the word *tana* reflects a meaning of "earth mother" and so again, taking earth as a symbolic term for mass, imply notions of the massive dark star. The word *penu* eventually came to be a generic term for deity in ancient India. Likewise, the stellar symbolism that characterizes sister goddesses in many ancient cultures through association with the Sirius stars (such as Isis and Nephthys in ancient Egypt) rests on precisely the same contrast, that between a bright, luminous star and a dark, massive one.

Of course light symbolism is also central to other ancient traditions we study. By around 2600 BCE in Egypt, it was the sun god Ra who rose to ascendance as patriarchy was seen to supplant matriarchy cross-culturally. It becomes clear in many ancient and modern religious traditions that concepts of light were closely associated with those of spirituality and knowledge, perhaps most obviously with the very term *enlightenment*. The Dogon priests flatly say that the nonmaterial universe is "of the nature of light," a curiously indirect statement (why not simply assert that the nonmaterial *is* light?), which upholds our outlook that nonmaterial energy in its native form may bear a closer kinship to magnetism than to light as we familiarly know it.

The relationship we perceive between the elephant god Ganesha and concepts that relate to light are also upheld in various ways. An ancient Egyptian word for light, *aba-t*, is formed from the same root *ab* that we link to various iconic attributes of Ganesha.[7] Symbolically the word reads "measure" ⌐⌐ "of place" ⌡, followed by the light glyph, 🜨. A light-year, of course, is the fundamental scientific unit of measure that applies to the microcosm, while the cubit serves what we see as a corresponding role to a light-year for material measurement in the ancient world. Meanwhile we have previously discussed the ways in which Ganesha's symbolism relates to the processes that bind, or accrue from the binding of electromagnetism, of which visible light is a form. These are most particularly reflected in the concept of angular inertia. In the mind-set of the Dogon cosmology, the reconfiguration of light energy into structures of mass occurs within the Second World of matter, a domain that corresponds conceptually to the ancient Egyptian underworld or Tuat. The essential Egyptian symbol that was used to represent the Tuat was the image of a rayed star ⊛, comparable to the Dogon egg-of-the-world figure—the same figure that we believe corresponds to the vectored rays of angular inertia.

The significance that the concept of light holds for Dogon culture is evidenced by the simple observation that the term *ogo* provides a phonetic root for the word *Dogon* itself, as well as for the priestly title of Hogon (*ogone*). The pertinence of light to the processes of material creation is reflected in a Dogon cosmological myth, where a character named Ogo plays the easily recognized role of light.[8] Any doubt regarding the symbolic relationship of Ogo to light should quickly resolve based purely on linguistic comparisons, attributes, and actions ascribed to Ogo in the Dogon myths and the explicit cosmological context in which his actions take place. We see direct correlation between the name Ogo and an ancient Egyptian word for light that Budge gives as *aakhu,* which also serves as the name of an Egyptian light god.[9] Symbolically, the Egyptian hieroglyphic word for light reads "that which" ⌐ "comes from" 🦅 "the source" ⊖, followed by the light glyph, 🔆. Griaule and Dieterlen describe Ogo as "agitated," reminding us that the Dogon word *ogo* means "quick" and so reflects one of the quintessential features of light. In fact, in the scientific view, light actually defines a cosmic standard for quickness, while Einstein treats the speed of light as a kind of fundamental speed limit. Ogo is cast in Dogon myth as having taken his origin in the nonmaterial realm attached to Amma's placenta, which corresponds to the nonmaterial universe, but with the desire to create a universe of his own. Comparable to that outlook, light energy is also framed scientifically as an underlying source of uncertain origin for our material universe. According to the Dogon myth, Ogo's impulsiveness causes him to break off a square piece of Amma's placenta, but that choice carries with it the unforeseen consequence of forever separating Ogo, a male character who now operates in a material frame, from his twin sister, a female who we infer is situated nonmaterially. This story line repeats a cosmological theme in which the nonmaterial universe is treated as feminine and the material as masculine. We also understand that in the symbolism of the ancient cosmology, a square is an icon of the material universe.

The cycle of scrolling energy, which seemingly rests on the dynamic of a dipole, brings the two siblings closer together and farther apart but never again reunites them. Ogo's action is characterized as a revolt, taken in opposition to a preexisting nonmaterial womb of life, and so the mind-set of the cosmology is one in which the formative seeds of material reality and consciousness reside ultimately with the nonmaterial. An Egyptian word for "revolt" is *aten,* whose glyphs read "that by which" ⌐ material ⌒ energy 〰 resists 🧍."[10] Ogo is credited with the formation of two energetic spirals (one nonmaterial, the other material) that turn in opposite directions, like the spindles of a papyrus scroll, and are said to be essential to the existence of life in the universe. This energy, which symbolically connects our universe to a placenta, is equated to an umbilical cord—the structure that serves a parallel and similarly vital function in the processes of biological creation.

We see a symbolic relationship to familiar attributes of electromagnetism associated with Ogo. As an example, the term *ogo* can refer to the concept of an axis, "symbol of the axis of the world,"[11] such as we know is defined by the perpendicular configuration of the dual energies of angular momentum, electricity, and magnetism. The scrolling energy that produces this axis is represented metaphorically in a variety of ways. We have said that it is the counterpart to a biological umbilical cord and symbolized by a scrolling papyrus, such as a Torah scroll in Judaism. It is also represented as the stem by which a piece of fruit, such as an apple, is attached to a tree, a connection that provides an energetic link to a source. The action of the scrolling energy is equated to the macrocosmic wheel or spiral that is specifically associated with the constellation of Orion.

In an aside that is made in *The Pale Fox* we are informed by Griaule and Dieterlen that Ogo had "brothers" who like Ogo were attached to Amma's formed placenta.[12] In keeping with that idea, we're aware that electromagnetism expresses itself in a number of forms that include visible light, heat, and various types of radiation. An Egyptian word for "brother" is *sen,* a cosmological term that can also mean "image" or "reflection," the same term that characterizes the nature of material creation for the Dogon. Budge cites a single-glyph spelling of an Egyptian word for "brother," formed from a figure that we take to imply the concept of the formation of or increasing mass 𓏏.[13] Budge includes explanatory glyphs with his glyph entry, which can be read to mean "translation" 𓏤 "of energy" 𓈖 . Budge also defines a phonetically related word, *sensen,* to mean "to be on brotherly terms with."[14]

He notes that an alternate term for the same concept is *heter*, a word that we know (based on the phonetics of the cosmology) implies the coming together of the energies of the two universes. Scientifically speaking, electromagnetism (light) is a joining of two energies that is associated with the formation of mass. Ogo is described by Griaule and Dieterlen as the "first being to develop his personality" in opposition to the creator deity Amma, "thus introducing psychological diversification into the universe."[15] This statement also seems sensible within the mind-set of the cosmology since the translational dynamic of Unity to Multiplicity that is said to be communicated through light energy applies similarly to concepts of consciousness.

In *The Cosmological Origins of Myth and Symbol* we discussed an episode of Dogon myth in which Ogo was said to measure out the universe in eight billion "steps."[16] We understood the term *step* as a key word, one that refers to an alternate method of measuring out cubits. As noted above, the ancient cubit can be alternately defined either as the length of a person's forearm or their average pace or step. Each physical method ultimately represents a unit of measure that is relative rather than precise, since each person's unique body dimensions will cause them to measure lengths that differ from person to person, and perhaps markedly so. In his seminal work on ancient units of measure, *Historical Metrology,* mechanical engineer and aviation pioneer A. E. Berriman cataloged many divergent lengths for a cubit among widespread ancient cultures. While the notion of a relative unit of measure may perhaps seem counterintuitive to modern sensibilities, where an imperative exists for ever-greater precision in measurement, it served an important purpose within the ancient cosmology since its primary usage was to plot the alignment geometry for a ritual shrine whose symbolism depended on relationships between figures, not their specific size. This same relative effect of difference in measure also pertains sensibly to the light vectors evoked through angular momentum, whose lengths progressively increase with shifting time frames. From this perspective, the seemingly odd circumstance of variously sized cubits takes on symbolic significance, in that it cues us to the relative natures of light and time frame, which might otherwise be mistaken as invariant constants. In regard to the Ogo myth we know that while the scope of the universe is much broader than eight billion cubits, one of the traditional figures that has often been cited for the radius of the visible universe is eight billion light-years, and so the matched numbers suggest a symbolic equivalence between the concepts of cubits and light-years. In support of that suggestion, Budge lists an ancient Egyptian term for cubit as *aakhu meh,* a compound word that literally means "light measure."[17]

Symbolic references to light also have pertinence to the civilizing plan that is associated with the ancient cosmology. As an example, the Dogon word *ogo* is a term for chieftaincy, which is the quality of being in charge, and so pertains to the Hogon priests, who are figures of expertise in Dogon society. Similarly, we see light symbolism reflected in ancient Egyptian concepts of chieftaincy. Prerequisite to an understanding of that, we know that the name of the Hebrew god of light, Yah, compares to an ancient Egyptian term for light that Budge gives as *aa.* Similarly, the Egyptian title of pharaoh is understood to derive from the words *per aa,* or perhaps literally "mouth of light," perhaps implying the authoritative influence of an enlightened voice. Concepts of light are also evident, as a convention of the cosmology, in the titles of priests. For example, a Dogon priest, or Hogon, is traditionally seen as the oldest and wisest member of a region and so also outwardly offers an enlightened voice. In keeping with concepts of electromagnetism that pertain to light scientifically, he is also associated with icons that represent the axis of the world.[18]

Virtually every ancient spiritual celebration is tagged to an effect of light. Countless shrines and temples of various cultures take their alignments from light, their architecture often specifically framed to emphasize a particular aspect or effect of it. Religious holidays traditionally fall at the equinoxes or at the solstices, calendar points that mark changes in the sun's annual north-south transit, along with the commensurate seasonal effects of light. The start of the solar year itself was marked in many ancient cultures by the heliacal rising of Sirius—the moment in the year when the brightest star in the night sky seems to come into closest proximity with the sun, our primary source of light.

11

LESSONS IN
SACRED GEOMETRY

LIKE THE DOGON, WHO prioritize them as a culture, we understand that the most direct interpretation of ancient cosmological symbolism lies with original forms. Over the span of millennia, a trend that we often observe is the slow evolution of original, basic forms into later, more complicated ones. This dynamic is perhaps most clearly evident in the practices of the *I Ching* (*yijing*), where the fairly direct symbolism of three-line trigrams are understood to have morphed over time into a set of more complex hexagrams. The *I Ching* ultimately spawned more than a thousand different interpretive applications, all arising from a single, fairly straightforward practice. In various cultures over time we see a similar evolution of the concept of the zodiac, whose components Marcel Griaule observes were all present as discrete elements of the Dogon cosmology. The same elements, those associated with Dogon granary shrine symbolism, are only known to have coalesced into a traditional zodiac in later ancient historical eras, by around 700 BCE. In the form that is most familiar to many Westerners, symbolism is applied to various stellar influences that are associated with a person's date and time of birth in relation to the positioning of stars and constellations. As with other cosmological elements, we see the concept of a zodiac applied somewhat differently in ancient China, where a progression of animals, also linked to constellations, relates symbolically to groupings of birth years, rather than to ranges of birth dates within a single year. Because the matching component elements seem to have produced independent versions of a full-fledged zodiac among a variety of cultures, we can surmise that a prezodiacal concept of the original cosmology might well have existed but then was only later implemented.

This same evolutionary trend from more basic to more complex forms also applies to concepts of sacred geometry, which make their earliest appearance in simple form, then over time almost literally blossom into the complex geometric tori and mandalas that often characterize the concept in the modern mind. In the cosmological tradition that we have been pursuing, sacred geometry begins with the very basic geometric alignment plan for a ritual shrine, in which we see scientific, practical, and symbolic purposes reflected all at once. We have said that in cultures such as Buddhism, learning to plot this geometrical alignment has traditionally been one of the first exercises introduced to new initiates, who may execute it in sand on a flat terrace and are expected to commit the fairly straightforward alignment sequence to memory. Archaically, the suggestion from the Dogon and the Buddhists is that initiates to the tradition measured out this same alignment configuration in an actual field. From the standpoint of practicality of implementation for the teachers, this was an exercise that required no special technology, tools, or supplies other than an empty field, a reasonably straight wooden stick, and a willing initiate. The alignment process depends as a prerequisite on the concept of a unit of measure, in this case the cubit, which also required no special supplies or technology to work its effect. We have said that, like all good cosmological constructs, the notion of a cubit carries more than one definition; as noted above, it can be measured as the length of a person's forearm (from the elbow to the tip of the middle finger) or as the average pace or step of a person.

This very same geometric alignment technique, with only minor differences, is an ancient and modern feature of Islam and is still used to determine the aligning of shrines and temples. Anciently, by some references, a plumb line could be used to confirm the vertical placement of the wooden stick. In ancient China the same geometry was seen to routinely define the ground plan of the earliest Chinese cities and shrines. A related shape comparable to the Chinese city plan, which is preserved in Egyptian hieroglyphic writing, defined an ancient Egyptian figure that Budge represents as the "town glyph." Appropriate to the ritual's significance in Buddhism, Adrian Snodgrass explains the intent behind the alignment geometry early in his book *The Symbolism of the Stupa*. He writes:

> The ritual of laying out the confines of the stupa to accord with the directions of space is the demarcation of an ordered, or cosmic space from out of the chaos of unlimited extension. . . . The significance of the ritual of demarcation lies in the fact that it is a mimesis of the measuring out of the cosmos from the dark and limitless Ocean of Universal Possibility. The ritual is a reenactment of this cosmic generation.[1]

In other words, as we have asserted, the alignment geometry (sacred geometry in its simplest, archaic form) is understood to replicate the dynamics by which space emerges during the processes of creation. So to the extent that we understand the archaic cosmology to have rested on scientifically reasonable concepts, we might expect to find important scientific clues regarding how space comes to be deployed couched within this geometry.

In keeping with this outlook, the grand symbol of the Dogon and Buddhist cosmologies is an aligned ritual shrine; in the Dogon tradition it was a granary and in Buddhism a stupa. In our view, these shrines represent more recent iterations of a singular, archaic symbolic form, in that they evoke the same set of geometric shapes in the same sequence, with the same symbolism attached. The shrines are related to two markedly similar symbolic cosmologies, defined largely by the same concepts and symbols but given in different languages. The conceptual plans of the two shrines ultimately differ from each other only at their final stage, where the Dogon structure culminates in a round base and squared roof, while the Buddhist shrine reverses those two shapes. In our view, this difference in form reflects yet another of many apparent reversals in symbolism that occurred cross-culturally, sometime midway through the approximately twelve-thousand-year half-cycle of the esoteric tradition. One consequence of the difference in implementation of the underlying plan is that it can produce structures that differ considerably in their outward appearance. Those differences may help account for the failure of so many researchers over a period of decades to recognize an association between the Dogon and Buddhist cosmologies.

While the plan of these shrines rests most immediately on the relationship of geometric shapes, at a somewhat subtler level of understanding the shapes can also be seen to reflect a progression that reflects dimensionality. This view begins with the central gnomon, or stick that is set vertically in the ground and around which an initial circle is drawn. The gnomon corresponds to the concept of a geometric point, a form that, geometrically speaking, foreshadows existence but has no actual dimensionality of its own. In the symbolic language of cosmology, the idea of a point can be expressed alternately as that of *place*. In the Dogon shrine model, the height of the shrine is given as ten cubits. The boundary of the circular base is then measured out as a radius in cubits—again for the Dogon, as ten cubits. That measurement, from the gnomon to what will become the boundary of a circle, constitutes a line, a one-dimensional figure that demonstrates the concept of length. The circle that the initiate then proceeds to draw around the gnomon takes its form in two dimensions, those of length and width. The shrine itself is completed in three dimensions as the two-dimensional plan is given height.

This alignment geometry is also directly associated with concepts of time, and less obviously with how the effects of time are evoked materially. For Einstein, time constitutes a fourth dimension, whereas the mind-set of ancient cosmology suggests that we should consider time as the first of four dimensions. Most obviously, the original circle of the shrine's geometric plan, with its central gnomon, constitutes a sundial, and so facilitates the measurement of time during daylight hours. In cultures such as ancient Egypt, the daylight hours (from sunrise to sunset) were associated with the material domain, while during the hours of the night the sun was said to descend into the nonmaterial realm of the Underworld. At night, time was often measured in relation to the progressive rising of constellations, which were used to mark the passing hours.

In the mind-set of the cosmology, the incremental motion of shadow in a sundial constitutes what characterizes material time as a *glimpse*; the fleshed-out constellation images correspond to the term *glance,* the term that characterizes time nonmaterially. As noted previously, a secondary result of the alignment geometry gives visibility to the seasons of the year through the daily plotting of an east-west aligned line. The placement of that line moves progressively northward and southward with the cycles of the solstices.

In *Seeking the Primordial* we came to understand that these two distinct time functions of the stupa alignment geometry also reflect two dynamics of time itself—an oscillation of time that associates with the nonmaterial universe and a vectored arrow of time that we all experience within the material universe. We know based on Einstein's outlook that space and time are interrelated concepts, so it makes complete sense that the same geometry that purports to illustrate how space emerges might also tell us something about the dynamics of time. As we understand them, both are by-products of the oscillation of a dipole with its positively and negatively charged centers that also induce energy to spin, thereby evoking angular momentum. As we have suggested, the material vector of time goes along with the effects of angular inertia and is situated perpendicularly to the plane of the spinning energy.

As Snodgrass explains the Buddhist outlook, sacred geometry delineates a sacred space, or *templum*, from chaotic space. He notes that in the original Latin, the word *templum* referred to "a sacred precinct set apart from the profane."[2] In the mind-set of the cosmology, the term *temple,* likely in keeping with the cosmological syllable *tem,* meaning "complete," refers to a place where the nonmaterial and material come together. Microcosmically speaking, this definition also refers to the aether unit, which corresponds to the almond-shaped vesica piscis that emerges as a consequence of the overlapping geometry of the two secondary circles of the stupa plan. Such structures, which can be thought of as taking form at the interface between the nonmaterial and material domains, are said to create the latticework of the material universe. We have said that these represent ripples on the surface of the waters and so effectively straddle the metaphoric domains of the two universes. The suggestion is that within these spaces an ultra-quick time frame persists, and so the rules of traditional physics break down. It is here where we believe entangled particles take on their seemingly "spooky" properties, which we also understand largely as natural effects of an ultra-quick time frame.

It is to this same overlapping space that archaic concepts of *deity* point. It was Adrian Snodgrass who noted that, from a Buddhist perspective, the concept of deity (*datu*) originally referred to aspects of material creation that are not subject to change, and the construct of the aether unit meets that definition. This is the same space that constitutes the "seat" of creation. For the Dogon, it sits at the center of the spinning energy of angular momentum, the "egg" of the egg-in-a-ball where, in the Dogon conception, Amma resides. It is the space formed when Amma's clavicles come together to "grasp," "hold firm," or "establish" the energies of creation. The very name Amma, of course, points us back to Ganesha and his goddess mother Sati/Parvati, whose symbolism also aligns with the aether unit. This is the yu seed, a term whose phonetics we said recalls that of Yah and the concept of light. Similarly, the Dogon syllable *ga* expresses concepts of temporality (past, present, and future). Taken together we evoke the notion of a Yuga, a cycle of energy whose influences rest on the relative slowing and quickening of time frame, as noted above.

Beyond defining the Dogon granary as aligned to the cardinal points, Griaule and Dieterlen do not provide us with an instructed method of alignment for the shrine; however, they do relate its specific dimensions and relay important details about the symbolism of the structure itself. This symbolism has multiple levels of significance. To begin with, the shrine constitutes a kind of student's primer in mathematics, with lessons first and foremost in practical geometry, counting, and measurement. Like a Buddhist stupa, features of the shrine play out in terms of relationships between a circle and a square and so reinforce a familiar cosmological theme—that of the nonmaterial and material coming together, expressed metaphorically in terms of the geometrical notion of squaring of a circle. Through its diameter, height, and number of staircase steps the granary plan defines ten as a familiar unit of magnitude. This is in keeping with the definition of ten-day weeks in ancient Egypt and ancient China. The number ten also reflects likely biological symbolism, that of ten fingers or toes, in much the same way that cubit-based units of measure take their definition from the human form, expressed in relation to fingers and palms and the stepping action of a person's legs. For initiates, each ten-cubit measure reinforced the basic mathematical concept of counting and helped to illustrate the concept of a staged sequence of events. We see this with the sequence of ordinal numbers that we know underlies such philosophies as Daoism in ancient China.

In relation to the solar system, the round base of the Dogon shrine, which takes the same configuration as an Egyptian sun glyph, represents the sun, while a circle that is inscribed within the square roof represents the moon. The reasonable implication is that the granary itself, which sits "between the sun and the moon," signifies the Earth. The same symbolism arguably applies to the Great Pyramid, whose form and symbolism mirrors numerous features of the Dogon granary structure and which is also understood to reflect aspects of the dimensionality of a hemisphere of the Earth. By incorporating both the figure of a circle and a square into its configuration (symbols of the nonmaterial and material universes), the granary associates itself with the concept of a temple, which we have said is defined as a place where the two universes come together. Symbolic elements of the granary reveal it to reflect aspects of both material and biological creation. These include its hemispheric shape (an Egyptian symbol for mass) and its association with the concept of Earth (a cosmological code word for mass or matter), as well as its known symbolism as an expanded womb (the biological correlate to mass).

At the center point of the aligned axis of the granary shrine rests a cup that holds two millet grains. For the Dogon, the millet grain constitutes the smallest seed. Its almond shape is reminiscent of the vesica piscis. On one level of understanding the grains correspond to the two energies whose interaction catalyzes the formation of matter. On another, they suggest the sets of female and male genes that must combine to initiate the processes of biological reproduction. Internally, the plan of the granary shrine defines eight chambers—four lower and four upper ones—one for each of the eight original varieties of seed of Dogon agriculture. Outwardly, the Dogon granary displays four flat pyramid-like faces, which are associated symbolically with four constellations that regulate the agricultural cycle. As noted above, the four ten-step staircases (one centered on each flat face of the shrine) are associated with categories and subcategories of plants and animals and so serve to illustrate the organizational concept of hierarchies.

The choice to associate phases of the agricultural cycle with constellations mirrors a key aspect of the Yuga Cycle concept, which similarly associates the eras of the descending and ascending segments of the Great Year with the full circle of constellations, whose rotation is conceptualized as the cycle of precession. The symbolic assignment accustoms us to conceiving cycles in terms of those same rotating constellations. Likewise the constellations are themselves often linked to animals whose symbolism also has significance for the cosmology. Such structures prepare us to apply that same outlook to the cycle of scrolling energy that defines the two universes in an even broader way.

The stupa alignment geometry measures time in an indirect way, in relation to shadows that are once removed from the apparent progressive motions of the sun. That dynamic reinforces the Dogon view of the structures of material creation as a kind of reflected image. Symbolically speaking, the sun is treated as an icon for the material universe, much as Sirius is often seen as an icon for the nonmaterial universe. For many cultures, the year itself began with the heliacal rising of Sirius, the moment when these symbols of the two universes make their annual appearance in closest proximity to each other. The directional motion of the shadow cast by the gnomon of a sundial can be seen to reflect the mode of linear time that we all experience materially, while the daily ritual of plotting of an east-west aligned line, as in the secondary stages of the alignment geometry, comes to mimic the oscillation of time that we believe characterizes the nonmaterial realm. In these ways, sacred geometry in its most basic and archaic form gives visibility to these processes and provides us with a conceptual tool for reconciling the two modes of time. When we take cosmological references in the metaphoric context of a Great Year, we see that emphasis is consistently placed on the significance of the equinoxes, which by ancient symbolic definition imply the idea of nonexistence coming into existence. An implication we draw from the dynamics of angular inertia is that the equinoxes represent a point of least resistance between the two domains. The vectors of energy that are evoked through angular momentum are subject to the least deflection at right angles to the spinning energy. If we were asked to paraphrase the core message that we see communicated across the ages through the forms of sacred geometry, it would be that the two modes of time (nonmaterial and material) seemingly come into reconciliation with each other at the equinoctial points.

12

NONMATERIAL TO MATERIAL TRANSLATION

A PERSISTENT THEME OF the ancient cosmology is how Unity becomes Multiplicity, how a single, grand, unified nonmaterial source expresses itself diversely in a material frame. Scientifically speaking, the idea of something that begins as nonmaterial but then somehow takes form materially may seem like an unorthodox proposition, but in truth it is consistent with mainstream thought in the field of astrophysics. The traditional paradigm for material creation rests on an understanding that, at quantum scales, matter exhibits behaviors that can be both wavelike and particle-like. Matter in its wavelike state, as we understand it, is virtually massless, a circumstance that, in keeping with Einstein's concept of relativity, implies that a wave must persist within the context of an ultra-quick time frame compared to ours. Quickness of time frame is the quality that, from our point of view, ultimately distinguishes an aetheric state of entanglement from the more familiar unentangled state of our everyday material universe. It is worth saying again that, from an outside perspective, entanglement takes on the attributes and properties of Unity.

The outlook is that an inherent translation occurs whenever energy moves across the boundary between the nonmaterial and material domains. The translation is effected by means of spinning energy that centers on the aether unit—where through the influence of angular impulse energy comes to be wrapped and configured—and by means of the vectoring effect of angular inertia. In that context, an Egyptian word for "boundary" is *atcher*, whose glyphs imply the idea of "that which is raised up and differentiated 𓊪 by the aether unit ⬭," followed by the distance glyph 𓊃[1]. It is through this process of translation that time frame slows and the concept of distance or space comes to have pertinence.

As represented in the ancient cosmology, the gross effect of this translational dynamic relates to how reality in an underlying wavelike form comes to be expressed as material particles. Conceptually, a body of wavelike energy is transmuted into a range of different vibrational frequencies of energy. We, in our material universe, perceive these vibrational frequencies as material structures. Similarly, and as what is represented as a parallel creational process, a grand unified but expressly nondeified consciousness comes to be individualized through the auspices of biology. The suggestion is that this grand consciousness also gives the outward appearance of Unity as a consequence of quickness of time frame in which it persists, much as we believe it does with entangled electrons, From the viewpoint of esoteric Buddhism, a final stage of enlightenment for an initiate comes with understanding that the diverse examples of individual consciousness we observe are ultimately all expressions of the same grand consciousness.

We have said that perhaps the most familiar real-world example we can point to of the dynamic of Multiplicity comes about when we shine white light through a crystal so as to evoke a rainbow of colors. In ancient traditions such as Buddhism, white is the color that is assigned to the nonmaterial realm, while green, for reasons that may not be immediately obvious, is often similarly associated with the material realm. Although on closer examination a rainbow is actually comprised of a continuum of progressive shades of color, our eye more typically perceives seven distinct bands of colors that correspond to different frequencies of light. A similar effect occurs with sound, where only seven frequencies along what is also a continuum of frequencies properly constitute the seven musical notes. We arguably see this same translational dynamic play out with seven vectors of energy that evoke the Dogon spiral of matter.

One ancient Egyptian word for "white" is *ubash*, whose symbols read "the spiral ⌇ of spiritual 🦅 divisions ▭ of light ⚏."[2] By one cosmological metaphor put forward by the Dogon priests, the evocation of the spiral has the effect of a sieve, comparable to the dispersive effect that a crystal has for white light. Our outlook is that the seven rays of the spiral coincide with the seven vectors of angular inertia. Symbolically speaking, we associate this type of vector with the Egyptian term for scepter and the geometry of angular momentum itself with the Egyptian sun-glyph shape ☉. In keeping with that outlook, another Egyptian word for "white" is *hetch-t,* and it can be written either with a single scepter glyph ⌇, or with two glyphs, as "scepter ⌇ of angular momentum ☉."[3] Of course we know that the scepter is also an icon of Ganesha's, whose symbolism we also believe reflects an aspect of angular momentum and the evocation of angular inertia. Similarly, an Egyptian word for color is *aun.* Symbolically, the glyphs of the word read "that which is ⌇ a vibration 🐦 of energy ∿," followed by a glyph that represents the stem of a fruit, 🎋.[4] The concept of a stem becomes another metaphor for vectored energy, in that it is a familiar structure that links the vitality of a living thing to its source. Yet another Egyptian word for color is *rit.* Conceptually, the glyphs of the word read, "the concept of the aether unit ⬭ and the existence ⎠⎠ of particles."[5]

Again our viewpoint is that the action of time in a nonmaterial context is that of an oscillation, reflecting an effect of the same action of a dipole that accompanies spinning energy. The material arrow of time that we perceive emerges as one of seven vectors of energy that are induced through angular inertia—more specifically, the vertical vector that sits perpendicular to the center of spinning energy, which is the fourth of seven vectors. We have said that in various cosmological contexts, green is a color that comes to be associated with the material realm. We know that green is also the fourth of the seven colors of the rainbow and so holds a position at the center of the progression of those colors—a place that is comparable to that of the vertical vector of angular inertia. More essentially, any cosmological reference to the number four should suggest to us a relationship to dimensionality, knowing that, if we count time as a dimension, our material reality rests fundamentally on a four-dimensional perspective. Looked at from a symbolic viewpoint in which dimensionality is expressed in relation to the concept of reeds and similarly in symbolic language through the notion of reeds ⌡, it seems sensible that an Egyptian word for "green" is given as *akh.* This is a term that can also be taken as a phonetic root of the word *ogo/aakhu,* which means "light." When we substitute concepts for glyphs, the word for "green" reads, "comes 🦅 from the source ⊜ as reeds 🌱."[6]

Another of the recurring metaphors that conveys the dynamic of Unity to Multiplicity is drawn from the natural water cycle. It compares the wavelike nonmaterial universe to a body of water and the many material forms evoked from it to raindrops. One implication of the metaphor is that a translational effect of diversification may be inherent to crossing the boundary between those two domains. The aether unit, or point of overlap between the geometries of the two domains, is described as a gateway between them. We see symbolic suggestion of this dynamic in an ancient Egyptian word, *agep,* which can mean both "cloud" and "rain."[7] Interpretation of these words rests with the view that a sanctuary or shrine, akin to the aether unit, is a place where the nonmaterial and material domains come together in the form of energy that spins. The glyphs of the word for cloud read, "The ⌡ shrine ⬠ of space ▢ evokes ▢ rain drops ⫴ as it is overturned ⟊ materially ⊙." Similarly, the word for "rain" reads, "That which ⌡ the shrine ⬠ of space ▢ encompasses ▭."

We suggested in *Seeking the Primordial* that the dynamic of Unity to Multiplicity carries with it implications both of geometry and dimensionality. In that context, a geometric point, which is effectively nondimensional (one-dimensional, if we count time as the first of the dimensions), becomes symbolic of Unity. This harkens to the symbolism of Amma, who the Dogon say resides in the central "egg" of the egg-in-a-ball figure ⊙. A geometric line, which takes its definition from two points, is representative of the concept of the measure of length in one dimension. This is the same concept of dimensional measure that pertains to the cubit. A square is an example of a figure of two dimensions in much the same way that a cube is representative of three dimensions. So it seems sensible that the Egyptian bent-arm glyph ⌐⌐⌐, which defines the concept of a cubit, is also the Egyptian figure for the numeral one.[8] Similarly an Egyptian prefix that implies the concept of "many" is written with the kite glyph 𓅷, which we have previously associated with the cosmological concept "comes to be." A comparable mind-set of translation across a boundary is conveyed in an Egyptian word for the number seven, which Budge gives as *sefekh*. Symbolically the glyphs of the word read, "Translated ⌐ or transmitted ∿ through the source ⊜," followed by seven dashed lines, which represent the Egyptian number seven.[9]

Looked at from the perspective of this one-to-many dynamic, if we entertain Samkhya's outlook that cosmological concepts, as the essential "seeds" of knowledge, ultimately reside with the nonmaterial, then this same distributive dynamic potentially explains the clusters of meanings that characterize these concepts in the ancient languages we have studied. It is reasonable to see these either as the product of, or symbolic of, the same transitional effect. The admittedly controversial Dogon outlook is that, during eras when the time frames of the two universes came approximately into parity with each other, a nonmaterial consciousness was able to take direct instructional action in a material frame. Our inference is that these clusters of cosmological word meanings stem from concepts that originate on the nonmaterial side of the boundary. We often see these same clustered meanings turn up in other settings where there is an arguable connection to the nonmaterial, such as in the context of vivid dreams or in eyewitness reports of UFO encounters.

On occasion during our comparative studies we have correlated words of two or more cultures whose meanings seem to properly align but whose pronunciations differ. These may relate to what the Hebrew language defines as dual consonants—letters that can reflect more than one pronunciation in different situations. The suggestion is that the dual consonants were original to the cosmological tradition but ultimately survived in some cultures with only one of the dual pronunciations. Examples of this are seen with the letters *peh/feh, kaf/chaf, dahlet/ thallet*, or *tav/thav*. In this context, the term *pil* or *fil*, whose clustered meanings seemingly pertain to Ganesha, would be understood to be the same word. Similarly we find archaic Egyptian terms such as the word *get* pronounced in later eras of the tradition as *het*, or arguably even as *chait* in the Fertile Crescent region. Kabbalism purports that there are seven such Hebrew letters that exhibit this property of doubled or variant pronunciation. From the view of retrospective analysis based on language, these variant pronunciations are a great help to researchers such as me, when attempting to track the likely flow of influences of the cosmological tradition from era to era or region to region. Taken in the context of what is arguably a designed symbolic system whose meanings were seemingly aimed at a distantly future audience, the introduction of such a feature as being trackable would seem to be deliberate. If the relationship to seven letters were part of a deliberate pedagogic choice, then a symbolic relationship to other expressions of the numerology of seven in the tradition, such as the nonmaterial/material expression of multiplicity, also might be inferred.

13

SELF-CONFIRMATION
OF MEANING

AS THE SAMKHYA AND Dogon philosophies frame things, the roots of esoteric knowledge reside in the nonmaterial side and are ultimately communicated, or translated in various ways, from the nonmaterial to the
material domain. More essentially, the "seeds" of reality itself are said
to be retained by the nonmaterial universe, and what we in the material universe perceive as reality actually constitutes a kind of reflected
image. This perspective recalls Plato's famous allegory of shadows on a
cave wall. Since the nonmaterial realm is understood to be of the nature
of light, the process of inter-universe communication is identified by
the term *enlightenment*. In keeping with that outlook, Budge lists an
ancient Egyptian term, *nems,* for "enlighten," which is given symbolically as "energetic 〰 concept ⌀ translated ⌷ materially" ⊙.[1]
From the perspective of Samkhya, the nonmaterial realm is described as
having "perfect knowledge, but an inability to act," while the material
universe is said to be endowed with "imperfect knowledge, but with full
ability to act." Because of that, Samkhya attests that routine attempts
are made by the nonmaterial to communicate knowledge to (or induce
action by) individuals on the material side. Such attempts can take the
form of vivid dreams, synchronicities, the odd behavior of animals, divination, clairvoyance, or prophecy and other effects that often fall under
the broad umbrella of the term *paranormal*. Because these effects can

be subtle ones, a legitimate question arises regarding how to distinguish a legitimately esoteric experience from other everyday experiences. For example, how should an individual differentiate between a simple everyday coincidence—as even the authors of the Samkhya philosophy admit happens—and one that might be esoterically meaningful? An instructive answer to that question lies with the concept of *self-confirmation of meaning*, the idea that, as an ongoing feature of the esoteric tradition, we can expect meaningful communication to be expressed in ways that affirm its own meanings. It is this feature of the esoteric tradition that often allows us to demonstrate, and thereby effectively confirm an interpretation in various ancient contexts. In our experience, such affirmation of meaning can take a variety of forms.

As a practical example of how this feature of the cosmology can work, we know from the Dogon priests that the plan of their stupa-like granary shrine evokes a squared roof that measures 8 cubits by 8 cubits, or 64 square cubits. The structure also sits on a circular base whose circumference, when calculated as $2\pi r$, using 3.2 as an approximation of π, also measures 64 cubits. The repetition of this same value in two geometric contexts within the same shrine plan argues that these are not just incidental measures, but rather careful and deliberate ones and so confirm the use as an intentional one. The effect of the shrine plan is to reconcile a square with a circle, which is one of the abiding metaphors of the cosmological tradition. Likewise, knowing that these matching mathematical measures still survive with the Dogon plan distinguishes their shrine as a likely original form of the symbolic tradition. The self-confirming measures also testify to the accuracy of transmission of the Dogon shrine plan. Moreover, the matching values serve to cross-check the specific dimensions of the shrine as Dogon cultural memory has preserved them and so lend confidence to the correctness of symbolic meanings that the Dogon directly assign to them.

In my third book, *The Cosmological Origins of Myth and Symbol,* it was this same quality of corresponding measures that brought Barnard's

Loop (a very faint spiraling birthplace of stars centered on Orion's Belt) to our attention. A Dogon mythological character named Ogo, whose name means "quick," and who plays the role of light in the Dogon cosmological myths, is said to have "measured out the universe in 8 billion 'steps.'" The perceived relationship of Ogo to concepts of light is upheld by the phonetically similar Egyptian word for light, *aakhu,* which is also the name of an Egyptian light god. We know based on general familiarity with the cosmology that the term *step* is one of the definitions of a cubit. While it's certain that the measure of our universe is much, much broader than 8 billion cubits, a figure of 8 billion light-years is actually one that is often cited as a dimension (radius) of the visible universe. Similarly, an ancient Egyptian word for "cubit," which Budge gives as aakhu meh, takes the familiar form of a compound word in English, one that can be literally translated as "light measure." The term again incorporates the Egyptian word *aakhu* that we previously correlated to Ogo's name and so affirms an association between the measure of a cubit and the Dogon character. We know that researchers inferred the size of an Egyptian cubit from the physical dimensions of the Great Pyramid, which measures 440 cubits per side of its square base, and 280 cubits in height.

My fellow researcher Robert Bauval sees an association between the three largest pyramids at Giza and the stars of Orion's Belt, while scientific citations for Barnard's Loop list its dimensions as 440 by 280 light-years. We take the corresponding figures in cubits and light-years not as coincidental, but rather as confirming measures. As further affirmation of the outlook, we learn that the Dogon also place large stones on a plateau to represent stars that have significance in their cosmology, specifically including the Orion's Belt stars. The stated purpose of these placements is to point us to what they define as the Chariot of Orion—a term that Marcel Griaule's team took to refer to the full constellation of Orion. However, when time-lapse photography is used to image the very faint light that is emitted by Barnard's Loop, the figure that emerges is a spiral that resembles the wheel of a chariot in which Orion the Hunter stands. Scientifically speaking, Barnard's Loop is categorized as a stellar bubble. Its scientifically surmised method of formation aligns quite closely with that of the Dogon spiral of matter. The identity of these various references upholds an as-above-so-below correlation that the Dogon priests make between their tiny microcosmic spiral of matter and this macrocosmic spiral.

In *China's Cosmological Prehistory,* we discussed a group of quasi-mythical Chinese emperors and the surviving story lines that relate their personal histories and accomplishments.[2] In China, long gaps of time elapsed between ancient events and the first surviving citations that refer to them. This gap between origin and interpretation creates huge uncertainties and has fueled persistent academic disputes among Sinologists over the intentions and meanings of these references. Our approach to bridging that gap in understanding was aided by features of ancient language. In general, we can observe that the further back in time our historical perspective shifts, the more commonality of language we seem to encounter. Likewise, there is a tendency for highly significant words to remain in a language for very long periods of time—linguists refer to these as ultraconserved words. Moreover, we can demonstrate common formulation, symbology, meaning, and method of interpretation for ancient Egyptian and ancient Chinese hieroglyphic words. So one recourse when writing the book was to compare key references given in the Chinese emperor myths to ancient Egyptian word definitions. We discovered that, for any given emperor, the attributes, events, and actions that are recounted as having been significant to the emperor's life played out as homonyms of each other in Budge's *An Egyptian Hieroglyph Dictionary,* with the emperor's name as their common phonetic root.

As we observed this to be the case, our impulse was to logically conclude that the emperor in question must have been a mythical, rather than a historical, personage, since it seemed unthinkable that such references could just incidentally come together as the proper history of any real-life figure. However, as we have gained experience with how nonmaterial meanings seem to express themselves materially, there is a perspective from which real events, looked at in retrospect, might take on some of the same contours as mythical ones. The Chinese myths gravitate around the same phonetic clusters of meanings as the cosmological words that we correlate between cultures. The use of related words from these clusters in any given myth provides a degree of self-confirmation of meaning for the myth itself and again testifies to its correct transmission, down through the generations.

Perhaps more significantly, we also see self-confirmation of meaning as a pivotal feature of the Egyptian hieroglyphic language itself. Early in my process I noticed that some thirty Dogon drawings displayed similar shapes and symbolism as certain Egyptian glyphs. That observation led me to explore how those shapes, whose well-defined meanings I firmly understood, might relate to the definitions of ancient Egyptian words where the corresponding glyph shapes appeared. Through that process I came to see that the related concepts, as the Dogon defined them, also gave an intelligible meaning to those Egyptian words. This implied that there might be a conceptual aspect to the glyphs that went beyond their commonly accepted phonetics. If I had embarked on these comparisons with perfect foreknowledge, I would have begun my comparisons with the Egyptian word *met,* meaning "week." The word is written with only two glyphs—the sun glyph ⊙, which can imply the concept of a day, and the Egyptian number ten ∏.[3] To my symbolic sensibilities the glyphs reflected the concept of "ten days," which, as it turns out, was the actual definition of an ancient Egyptian week. In effect, the glyphs of the word not only self-confirmed its meaning, they precisely relayed its definition. An ancient Chinese hieroglyphic word for "week" is formulated in precisely the same way and so demonstrates a kind of core commonality for the two written languages. Having observed this one example of self-confirmation of meaning expressed in the glyph structure of an Egyptian word, my natural impulse was to wonder whether the glyphs of other Egyptian hieroglyphic words might also work in this same way. I quickly could see that they all seemed to work that way.

We know that the archaic cosmology was originally an oral tradition, and symbolic written language is understood to have been a much later development. A number of cultures, such as the ancient Egyptians, deemed their written languages to have been gifts from their deities. Notwithstanding that view, we can see significant overlap between drawings that illustrate the oral Dogon cosmology and written glyphs, a circumstance that suggests that written symbolic language evolved as a late feature of that tradition. Moreover, the Egyptian hieroglyphic words reflect the very same quality of self-confirmation of meaning that characterizes the oral cosmology itself. These written words also consistently reflect the same clustered meanings that we associate with root concepts of the oral cosmology, which again argues for a common origin. The phonetics of Dogon language represents an entry point to self-confirmed meaning in the ancient cosmological tradition. To the extent that each syllable aligns with a cosmological concept, we understand that the simple pronunciation of a Dogon word can telegraph its cosmological meaning. It is in this context that we infer that the name of the creator god of Dogon mythology, Amma, reflects actions that initiate the processes of material and biological creation. The syllable *am* can represent the concept of "knowledge," which in the biblical sense implies procreation, while the syllable *ma* refers to an act of perception, the act that initiates an increase in mass for a primordial wave.

For this same reason, we find other evidence of that phonetics in Hinduism. Knowing that Ganesha's eight incarnations align with stages of the po pilu, and that Amma was the affectionate nickname by which Ganesha referred to his mother, Sati, we can infer that the self-same phonetics must have underlaid the Samkhya tradition in India at the time of the emergence of the Hindu tradition. As an example, the name Sati also combines two cosmological syllables and so conforms to the same phonetic convention as the Dogon name Amma. The syllable *sa* refers to the concept of "knowledge," and *ti* can refer to the concept of "time." Our outlook is that Sati's symbolism is to the action of virtual particles that form the aether unit, the gateway structure whose effect rests on the shaping of energy and slowing of time frame. It is easy to see that her consort Siva's name also conforms to this same linguistic convention. The syllable *si* refers to the "nonmaterial," and *va* refers to the concept of a "voice." If we endorse the Dogon view that the spiral of matter constitutes the Word of matter, then the corresponding notion of mass and time frame, which grows stronger or weaker as energy scrolls, represents its voice. These are the processes of material creation that differentiate particles from waves and so mimic the effect of a sieve, a word that we associate with the process based on a Dogon cosmological metaphor.

In my experience, interpretation of symbolism for vivid dreams often rests with this very same principle of self-confirmation of meaning. If we endorse Samkhya's assertion that vivid dreams can be one of the conduits for communication of knowledge from the nonmaterial domain, then we can imagine that they might be subject to the same inherent translational effect between the nonmaterial and material domains that we experience in other contexts—that of unity to multiplicity. In these cases, Budge's hieroglyphic dictionary provides us with a useful tool for triangulating on likely interpretations for dream words and images. Of course, again, an implication of this view is that the related dream concept must take its origin within the nonmaterial domain. From the perspective of the Egyptian hieroglyphic language, we have said that the clustered meanings play out as homonyms—words that are pronounced alike but carry different definitions. For the Dogon, where words are spoken but never written, the traditional concept of a homonym doesn't exist, and so these clusters of meanings take expression as multiple definitions of a single word. In practice, Geneviève Calame-Griaule often distinguishes separate word entries of the *Dictionnaire*

Dogon for broadly differing definitions of the same phonetic word. I have often seen my own vivid dreams, and those of others, expressed through key words, images, and impressions whose names also play out as homonyms of a single Egyptian cosmological term. The same effect is often observed with UFO encounters, as documented in psychiatric interviews with witnesses by professional researchers such as John Mack and Budd Hopkins. An intuitive example involved multiple witnesses to a late night abduction in Manhattan who each, after a period of weeks, independently recalled images of the abductee on a beach holding a dead fish. According to Budge, an Egyptian word, *maut,* for "dead fish" can also refer to "part of a story to be remembered." Many times there will be more than one thread of action within the dream whose Egyptian words coincide in a given pronunciation, and which therefore also gives the impression of self-confirmation. It is these coinciding references that provide us with a likely interpretation of meaning for the dream.

Synchronicities are another interesting case, and perhaps a more difficult one to affirm, in terms of esoteric significance. The experiencer is in the position of having to find ways to differentiate between a synchronous experience and everyday coincidences. For my own purposes, when the same seemingly odd reference turns up in unrelated contexts, my learned response is to explore whether it could possibly have pertinence to my studies. If the reference comes in the form of a word, then my first impulse is to check for entries for that word in the Egyptian hieroglyphic and Dogon dictionaries. Again, Egyptian hieroglyphic words provide an interpretive advantage in that when we substitute concepts for glyphs, the various spellings of the word can reflect specific nuances of meaning. There is a similar advantage to exploring words in Geneviève Calame-Griaule's *Dictionnaire Dogon,* which provides extended definitions for words formed from a given root. The dictionary also often notes alternate phonetics for a term—which makes correlations to words of other languages easier—and entries may also include extended commentary when the word is understood to have cosmological significance. If the seemingly synchronistic reference that I have experienced involves a concept that's unfamiliar to me, then I often search the indexes of various of my primary sources, hoping to get a consensus view of how researchers I most respect understand the concept. When more than one seemingly synchronous event involves a specific phonetic root or a specific shape, then comparisons to ancient languages or symbols are often helpful.

For the average person who might not partake in the types of comparative studies I do, confirming a synchronous event might be more of a subjective process. Simply allowing the possibility that a coincidence might have meaning can be a first step toward a better understanding of the esoteric.

For the purposes of my own work, I have learned to act on certain situations that have previously shown themselves to be esoterically pertinent and can now often be triggered by just a single odd circumstance. As an example, I often receive questions from complete strangers that may at first seem tangential to my cosmological studies but that I now make a point to follow up on. There has been more than one occasion when my choice to pursue the question has led to material that proves to be productive to my own work. For example, my book *The Mystery of Skara Brae* took form from just such a question, where the stranger inquired by email whether I felt there could be ancient Egyptian influences reflected at the Neolithic Scottish farming village of Skara Brae. At the time, I had virtually no knowledge of Neolithic Scotland and therefore might have been reluctant to divert my attention and energy to the topic. However, the choice to sincerely explore the question opened up a book's worth of material to me in a very short period of time. A somewhat different set of synchronicities catalyzed another of my books, *Point of Origin*. These emerged one July morning when I had been researching half a dozen different questions on ancient language for various friends or colleagues. On this particular morning, six threads of research all inexplicably resolved based on word entries drawn from the same column and page of Budge's Egyptian hieroglyphic dictionary. A seventh word from that same column demonstrated a phonetic identity between an Egyptian word for "pillar" and another for the concept of an "embrace," a metaphoric act that is pivotal to an understanding of the energetic dynamics of the cosmology. From there I came to see that various Egyptian words for "embrace" were phonetic matches for numerous other important terms of cosmology. An awareness of that symbolism offered me a new outlook on the somewhat coldly rendered embrace depicted on one of the Gobekli Tepe pillars, expressed by two amorphous carved arms and hands that extend down the sides of the pillar with bent fingers that seemingly embrace the pillar's narrow front. Those insights opened one of the doors for my book, the material for which again emerged over the course of only a few months.

Taking a broader perspective on the concept of cross-confirmation of meaning, we could say that it represents one of the first principles on which the successful study of comparative cosmology itself is based, in that the act of anchoring an interpretation often ultimately requires it. Because self-confirmation is a feature that is objectively observable in many different aspects of the cosmology, the mere fact of it demonstrates intention, and so upholds our understanding of the symbolic tradition as an instructed plan. In most cases, we require more than just a superficial view of comparative elements before their self-confirming aspects become evident—in other words, they may not be immediately apparent to the casual viewer, only to those who take the time to examine a situation a little more thoroughly. Precisely because of this, these subtle self-confirming elements are often among the first casualties of any effort at translation or reinterpretation and so tend not to survive in later forms; hence, the Dogon societal imperative to preserve original forms.

14

DYNAMIC OF THE INITIATE AND INFORMANT

AS IS TRUE FOR many aspects of the ancient cosmology, the dynamics of initiation in the esoteric tradition bear careful examination from a number of different viewpoints. Outwardly it may appear that the focus of the esoteric tradition is on protecting a closely held body of secret information from outside view. This might lead us to think of initiates and priests as a privileged class, and so see the tradition as elitist. However, a more nuanced perspective on the tradition suggests that it is more about preserving and conveying intact a body of universal knowledge to some future generation. Seen correctly, just as we understand that the esoteric tradition itself is open to anyone who sincerely pursues it, the knowledge it preserves is information that properly belongs to everyone, not merely to a few.

With regard to the esoteric tradition, Dogon society depends on the collective expertise of their most knowledgeable elders for the correct interpretation and transmission of a complex body of knowledge to the next generation. Whenever questions of a subtle or controversial nature arise, the outcome ultimately turns on a consensus view of these elders. One important effect of Marcel Griaule's study and eventual initiation into the Dogon esoteric tradition was that it tested and affirmed that Dogon esotericism was truly open to any person, even a foreign, Western-educated outsider. So unusual was this circumstance (an outsider in sincere pursuit of inner Dogon knowledge) that Griaule's priestly informant felt that special permission would be required from this council of Dogon elders before he could proceed with Griaule's initiation. The assignment of the blind Dogon priest Ogotemmeli as Griaule's informant was, of itself, an interesting one, in that it implied that the old hunter understood his subject so thoroughly that he could trace proper sand drawings of all of its essential shapes, even without the benefit of eyesight.

We have mentioned that in *Conversations with Ogotemmeli* Griaule provides an overview of the cosmological mind-set that guided the period of his formal initiation. In the book he comments on the careful precautions that were taken to distance his discussions with his informant, the blind priest Ogotemmeli, from the curious ears of any uninitiated Dogon villagers. These ranged from seeking out private places for intimate discussion, speaking in hushed tones when in the presence of any distant tribespeople who might overhear, to speaking under the cover of natural distracting noises of the village, such as the crowing of roosters. Also mentioned were the fines that could be levied against Ogotemmeli if the studies were deemed not to have been conducted with a sufficient degree of privacy. Griaule also writes about various objects of Dogon life that were adapted by Ogotemmeli as visual aids to illustrate these talks. The mere fact that everyday objects lent themselves so readily to use as illustrative aids underscores how closely integrated the artifacts of Dogon life are with their symbolic cosmology.

In other cultures, such as the Maori of New Zealand, there is evidence of two modes of esoteric initiation—one whose focus is on symbolic cosmology and the other on occult practices. Among the Maori, those who study cosmology are expressly forbidden from also studying the occult. Maori sources describe initiation into occult practices as culminating with a requirement that the student actually demonstrate the ability to kill a person using occult techniques. By contrast, Griaule and Dieterlen make little hint of any similarly dark occult practices in Dogon society—in fact, Geneviève Calame-Griaule includes no word entry for the concepts of "occult" or "magic" in her *Dictionnaire Dogon*. Such concepts clearly did exist for the ancient Egyptians, and in that context Budge lists the word *stekhi,* which he defines to mean "malign magic."[1] We could argue that the word is adapted from the phonetic root *heq* or *hek*, which for Budge refers to the concept of inflicting or diminishing pain.[2] The Egyptian word *heka* is often cited as a reference to magic of the occult. Griaule and Dieterlen consistently represent the Dogon priestly tradition as one that upholds a high, yet studiously non-judgmental and unmalicious ethic.

Looked at cosmologically, there are additional motives and intentions that we can infer for the esoteric tradition, and these express themselves in much the same way that the dynamics of matter do, in that we see them play out symbolically on what we might think of as upward scales. To put these motives into context, we first need to expand our outlook to accept that broader societal constructs of the Dogon and others, such as the esoteric tradition itself, can themselves also be seen as symbolic representations. In other words, we need to allow that the dynamic of the esoteric tradition may take the form it does in part because it replicates a more essential dynamic of creation. From this perspective, the relationship between the student of cosmology, who knows relatively little, and his or her informant, who seemingly knows everything, can be seen to mimic the one that is described for the nonmaterial and material universes. The informant plays the role of the fully ascended nonmaterial universe, which is said to have perfect knowledge but is somehow inhibited from taking action. The initiate stands in the place of the material universe, with imperfect knowledge but full ability to take action. And so the student is obliged to continue to instigate the next action; more specifically, to formulate and ask the next question, whose answer will further his or her own initiated status. When that question has been posed by the student and subsequently determined by the informant to be an appropriate one, the informant is then required by the tradition to respond to the question truthfully. The entire interaction is one that is familiar to some classes of psychic mediums who seemingly draw their information from a non-material source through a similar dynamic of inquiry and answer.

A thoughtful person might be tempted to ask why the ancient instructional dynamic takes this somewhat awkward form of question then answer, rather than the more straightforward mode of direct instruction that we might expect of a modern classroom. Of course, the immediate response lies with the outlook just previously discussed—that it replicates the dynamic by which Samkhya tells us knowledge is conveyed between the nonmaterial and material universes. However, with a sincere candidate this choice undercuts any natural resistance the person might have to receiving instruction (the tugged wild-grass effect), since it essentially puts the initiate in the driver's seat. It also requires the student to take ownership of his or her own esoteric education. This is the same dynamic that we see expressed in energy healing, where the source of beneficial health effects is understood to reside ulti-mately with the nonmaterial but is induced to work its effect materi-ally. In other words, the dynamic between an initiate and an informant in the esoteric tradition accustoms the student to the mode in which knowledge is said to express itself between the universes and is commu-nicated in the same way that it is ostensibly conveyed to a person who is spiritually sensitive, whether a shaman or a clairvoyant.

Another effect of a question-and-answer instructional dynamic that is driven by questions posed by the initiate is that it requires students to think deeply about what they have already learned, contemplate its implications, and then carefully formulate the next follow-on question. That process reflects an effective definition of the concept of thought-ful discrimination. The ethics of the informant's response, which call either for either a response of silence or a truthful answer, provide a mechanism by which the teacher is able, through silence, to subtly guide the student away from wrongful thinking and on to the next proper question, and also away from any line of inquiry that might seem overly ambitious. By this method, progress for an initiate is slow—arguably much slower than it would be with the type of direct instruction most modern students receive in a classroom. However, it is also one in which the successful candidate has the luxury of time to truly master the subject, a goal that is one of the top Dogon societal imperatives.

To the modern eye, the dynamics of Dogon society might outwardly resemble an inherently sexist tradition, in that strict divisions of labor are practiced that often tend to fall along male/female lines. As examples, the women traditionally are the spinners of thread, while the men are weavers of cloth; the men plow the fields, while the women and children sow the seeds and are instrumental in the harvesting of crops. However, as a model for an early society where the establishment of families and raising of children were likely among the primary existential concerns, these same divisions of labor seem eminently practical. One effect of the work assignments is that they leave the women free to attend to many of the unique duties of a mother (a rationale that could again be represented either as sexist or as practical). Similarly, men and women are segregated from each other in their study of cosmological subjects—the women study with female elders and the men with male ones—but at root, a path to esoteric knowledge is ultimately open to any person. From a broader view of things, the archaic tradition as we understand it was matriarchal, with esoteric knowledge passed orally from generation to generation by female yoginis. Knowing that fertility was a primary focus of the early tradition, it seems sensible that Dogon women still receive their instruction from female elders. In keeping with other broad symbolic reversals of the tradition that we observe, the later tradition came to be patriarchal, with emphasis on the male Hogon as the public face of esoteric knowledge.

As in Judaism, where the Cohane are traditionally seen as a priestly class, so is Dogon society separated into family lineages, each associated with one of eight mythical ancestors who, among other roles, serve as conceptual patriarchs for each family. Again like Judaism, societal responsibilities are also divided according to a religious hierarchy, in that certain functions are reserved for the priesthood. Such restrictions are perhaps most obviously reflected in concepts of tapu (taboo), which often pertain to instances of inadvertent defilement and ultimate restoration of ritual purity to a person, a location, or an object. In Judaism, similar tapu rituals survive, such as the ritual washing of a congregant's hands after leaving a cemetery. In the esoteric tradition, these same concerns also pertain to potential transgressions of propriety and secrecy.

The motive for secrecy in the esoteric tradition is another concern that a thoughtful person might come to consider. If the ethic of the tradition is that inner knowledge is ultimately open to any person who sincerely pursues it, then why bother to shield that knowledge behind a veil of secrecy? The first sensible motive for secrecy as an aspect of instruction is one that the Dogon themselves promote, and that turns on a conviction that the psychology of humanity is such that intuitive ways must be found to draw people to esoteric knowledge. The very hint of secrecy tends to pique human interest and so effectively entices a person to want to know about a thing.

The second motive for secrecy comes out of a much longer-term requirement for sincerity among the class of initiates. As discussed in *Seeking the Primordial,* the dynamics of scrolling energy between the universes (essentially the dynamics of the Yuga Cycle in Buddhism) are such that, during a universe's most ascended/least material state, consciousness becomes essentially locked in and remains so for a period of thousands of years. Again it seems that what we first take as a metaphor of the Samkhya philosophy, which characterizes the nonmaterial as having perfect knowledge but with no ability to act, might be taken on a more literal level. Any person in this same condition, who is paralyzed but still fully conscious, would find it to be a horrific state—essentially one of being buried alive. That horror can be mitigated if some more capable person can be made aware of the condition, and who is then able to act on behalf of the incapable person. Anyone who suffers in this state, known as locked-in syndrome, would most desperately need a sincere confederate to take notice of them, demonstrate concern for their wishes, and take certain actions on their behalf. Secrecy provides the esoteric tradition with a vetting mechanism, by which sincere initiates can be distinguished from others who may be less dedicated or less sincere. In this way, the dynamic of the esoteric tradition mimics the action of a sieve, a tool the Dogon priests use to illustrate the cosmological concept of the separation of particles from waves.

There is also a perspective from which the secrecy that is so carefully practiced with the instructed civilizing plan could have what are essentially political overtones. It is a dynamic that we would expect to see if those who provided the instruction had also felt a need to hide the fact of that instruction from someone. This perspective is lent credence by episodes from the Old Testament, most obviously with the story of Moses on the mountain. Here a nonmaterial entity ostensibly takes an energetic material form and makes its very first proclamations against belief in earlier religious traditions—suggestive of the symbolic tradition we have been exploring. Likewise, knowing that, in cultures such as the Maori of New Zealand, the cosmological term *firstborn* can refer to the priestly class of the esoteric tradition, we might imagine that the phrase slaying of the firstborn implied someone taking deadly action against the priests of the tradition in the story of the Exodus from Egypt. So it is possible that an educational tradition whose dynamic was a studiously secret one, and whose net effect was to essentially hide the very fact of that instruction from outsiders, might have been promulgated in the context of a powerful rivalry. There are other ancient story lines and references to support this outlook. Most obviously, the overt Dogon tradition is that after a period of time their mythical teachers either chose to leave or were forced to leave—an event that the Dogon saw as a turning point for humanity.

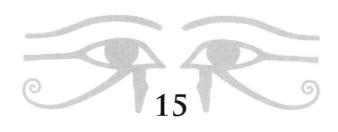

15

THE NATURE
OF WATER

THE HERMETIC PRINCIPLE, "AS above, so below" implies that there is an identity between processes of the macrocosm and those of the microcosm. We see the same basic principle described and illustrated more succinctly in the Dogon cosmology, and with specific examples cited—examples that consistently hold up to comparison from a scientific perspective. Once we accept this fundamental correspondence as a truthful principle, then there are important inferences that can be drawn from it—first and foremost that the dynamics of creation on any scale are likely to express themselves in forms that will seem familiar to us based on our everyday experiences. From there, the task of understanding any of these processes largely requires us to seek out the proper metaphor from within range of our own observations. The idea plays to our instincts—many a neophyte middle-school science student notices the similarity between the dynamic of an atom and that of a solar system, for example. Meanwhile, as if to underscore this idea, the plan of the ancient cosmology rests largely on a series of carefully framed metaphors whose very purpose is to frame subtle processes of creation in more familiar terms. This is the same technique we see employed when cosmological concepts are tagged symbolically to the characteristics of animals or objects found in an initiate's local environment. In that context, the careful integration of cosmological symbolism with objects and activities of daily life of an instructed culture can be taken as a gigantic hint about how we should approach perceiving the dynamics of creation—that, in our efforts to do that, we should look first to familiar dynamics for important insights.

One such parallel is found at the interface between the domains of a body of water and the air. Ancient symbolism suggests that the effects we observe at the surface of water offer us important clues about the likely nature of any interface that might exist between two energetic domains. Moreover, that same symbolism arguably rests on specific, recurrent metaphors to water, and these are references that we should surely not dismiss as incidental. We argued in the first book of this series, *The Science of the Dogon,* that Dogon descriptions of the Nummo as a perfect twin pair, intimately linked to water, might easily pertain to the dynamics of two hydrogen atoms as they are found in a water molecule. In fact on one level of understanding, the Dogon explicitly designate the Nummo as the Master of Water—in fact, the French title of Griaule's *Conversations with Ogotemmeli* is *Dieu d'Eau,* or "God of Water." We are flatly told by the Dogon priests that both the Nummo and the nonmaterial realm are "of the nature" of water. We find in various ancient languages that the cosmological phoneme *nu* consistently refers to notions of water and waves. It is the commonly shared view of modern science and the ancient cosmology that matter begins in wave-like form, which is also akin to water.

We know that Nummo is also the name that the Dogon assign to their mythical teachers, those who are credited with having been the conveyors of cosmological knowledge. The association with water makes sense in that it evokes the idea that esoteric knowledge is conveyed to humanity along with the energy that scrolls between the two universes. We find an ancient Egyptian word for water that reflects this idea symbolically. The word is *tehem,* which means "to water." A conceptual reading of the glyphs of the word can be interpreted as "materially ⌒ scrolling ⬚ knowledge 🦅," followed by the water glyph 〰.[1]

Familiar representations, such as in the Book of Genesis, of non-material energy as an initial glow of light that hovers over water effectively places any archaic discussion of the catalyzing principles of matter at the very interface between water and air—literally at the surface of water. In support of this view we find an Egyptian term for water given as *mu* and reflective of various symbolic representations in a number of word entries. One such entry Budge somewhat foggily interprets to mean "on the water of someone," which he believes to imply the notion of "dependent on someone." We see the purpose of the word as to specifically define a concept. Symbolically it reads, "Water's 〰 surface ▭" as a definition of "the concept of the face ☥ of the waters 〰."[2] Given all of that, it seems thoughtless not to look for parallels to the initiating processes of matter among the dynamics that govern that familiar interface. Moreover, we know that the dynamics of energy often reflect the quite similar dynamics of water, to the extent that parallels to water are commonly cited during discussions of energy. Since those of us who may be unfamiliar with the dynamics of energy do come into daily contact with water in its various forms, the motions, states, and modalities that define it make good metaphors for illustrating what occurs with energy. The Dogon themselves use such water-related concepts as *phase transitions* (water freezing to form ice, or water vapor condensing to produce water) as metaphors for transitional processes that occur energetically during the first stages in the formation of matter.

In the scientific view, primordial energy is given early expression through the emergence and dissipation of virtual particles. From this perspective, water becomes a symbol for the energetic source that gives material creation expression. This outlook is also supported by the ancient Egyptian word *net,* which happens also to be the name of Neith, the Egyptian goddess who wove matter. The symbols of the word read, "energy 〰 of matter ◠," and in our view define the trailing three-wave glyph that represents water 〰.[3] Virtual particles can be seen as conceptual counterparts to the rippling effects that we observe to emerge and dissipate on the surface of a body of water. We know that ripples on water are a by-product of the effects of wind, in very much the same way that ripples in energy can be a by-product of vibration. In fact, in the symbolic mind-set of the cosmology the very notion of wind becomes a metaphor for the concept of *vibration*. In another metaphor of the cosmological tradition, the cycle of scrolling energy that characterizes the two universes is overtly compared to the natural water cycle on Earth. We have said that perhaps the most familiar of the four-stage metaphors of the symbolic tradition, given as water, fire, wind, and earth, begins with the primordial element of water. In keeping with all of this, the action of virtual particles as they seemingly pop into and out of existence produces the same almond-shaped configuration that the Dogon represent as an initiating structure of matter—the same shape that they compare metaphorically to the tiny millet seed called the Yu. The shape also defines the vesica piscis, which is central both to Hermetic concepts of sacred geometry and to Buddhist conceptions of how the geometry of space and time emerge. So from numerous perspectives these water-like ripples represent a likely entry point for understanding the underlying energetic dynamics of matter.

As it turns out, well-known oscillatory dynamics of water, which begin as ripples and waves, are capable of producing many of the very same shapes that characterize Dogon discussions of the progressive stages of matter. Under the proper conditions, the action of waves can evoke swirling water in much the same way that energetic oscillation can cause energy to spin, reflecting the Dogon egg-in-a-ball configuration. Similarly, in a depth of water, this can produce shapes that compare to the Dogon spiraling egg-of-the-world. Moreover, the dynamics of swirling water (a water vortex) are virtually indistinguishable from those of a macrocosmic black hole, or from the tiny vortices that arguably define the Dogon spiral of matter, the po pilu.

In *Seeking the Primordial*, we inferred that the energetic dynamic of the universe is that of a dipole, whose effect is evoked by positive

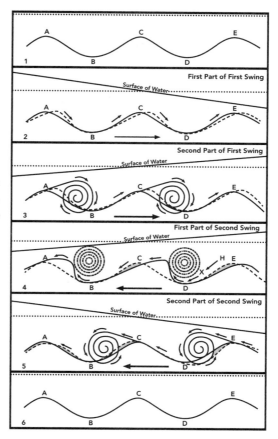

Fig. 15.1. Image from "The Origin of Ripples and Other Fantastic Fluid Experiments," by Hertha Marks Ayrton, Deep Sea News.

and negative energies that persistently oscillate together and then apart. We know that dipoles come in a variety of types that include magnetic dipoles, electrical dipoles—even gravitational dipoles. On upward scales of perception, the concept of a dipole seems to underlie other aspects of the universe including the dynamics of galaxy groups. We learn that in addition to the categories already mentioned, science also defines the concept of a molecular dipole and cites the water molecule as one of the quintessential examples. The polarity of a water molecule comes about because the oxygen atom carries a much greater negative electric charge than the two hydrogen atoms it links with. When hydrogen's single electron bonds with oxygen and shifts nearer to the oxygen atom, each hydrogen atom is left with an almost entirely positive charge. So each hydrogen-oxygen pairing oscillates like a dipole. An online science source clarifies the concept:

> Because of water's structure, the partially positive H's are on one side of the molecule and the partially negative O is on the other. Molecules like this with partial positive and negative ends are called polar molecules. . . .
>
> Finally, these partial charges interact the same way that full charges do—positive and negative are attracted toward one another. This is essentially what dipole-dipole forces are: the attraction between partial negative and positive charges.[4]

Although a water molecule is typically diagrammed as if the hydrogen atoms are configured separately from each other, in practice the molecule oscillates, in keeping with its dynamic as a dipole, so both its shape and the relationship among the three atoms (two hydrogen and one oxygen) can vary. There are also many anomalous aspects to the behavior of a water molecule, so much so that no coherent mathemati-

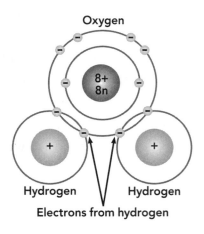

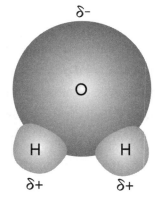

Fig. 15.2a. Electron shells in a water molecule.

Fig. 15.2b. Distribution of partial charges in a water molecule.

cal model has yet been successfully developed for it. As the configuration of atoms oscillates inward, the geometry of a water molecule takes on more than a passing resemblance to the geometry that is used to align a stupa shrine (see fig. 5.1 in chapter five).

Much as we have argued that the multiplicity of the material universe is a function of the density of mass, and that space expands hand in hand with the evocation of mass, so water becomes more diffuse as its density increases. We also know that a water molecule is capable of forming bonds with four other water molecules, and so the bonds themselves take on a suggestive relationship to the concept of dimensions. With all of this, a reasonable question arises as to whether, as with other comparative effects of the ancient cosmology, the structure of a water molecule could reflect another of the root dynamics that play out in parallel form on upward scales of observation.

We have noted that the atomic structure of a water molecule is comprised of three atoms, or particles. Within the creation tradition the notion of a particle can be expressed with the image of a clay pot ○, often represented as a container of water. Cosmological concepts of water are expressed in various ancient languages by the phonetic value *nu*. In keeping with this mode of expression, an ancient Egyptian term given by Budge for the concept of the *celestial waters* (or *primordial waters*) was *Nu*.[5] So it is interesting that one spelling of the word can be interpreted to mean "three particles ○ ○ ○ together ▭." The effect of the glyphs is to associate the specific molecular structure of water with the concept of *the waters* as a cosmological source. In other words, there may be a point at which we take the persistent metaphors to water in a more literal sense, in terms of how they pertain to the underlying dynamics of the universe. To the extent that we imagine the configuration of paired universes to reflect that of water molecules, the number seven also makes sense in that there are suggestions that seven is the minimum number of water molecules per ion pair to ensure biological activity in a body of water.[6]

16

UNITY AND THE DIMENSIONALITY OF NUMBERS

THE ARENA OF NUMBERS and numerology represents a place where the civilizing and cosmological aspects of the ancient creation tradition outwardly come together. We can easily infer that among the first concepts instructed to humanity was that of numbers, along with basic skills of counting, mathematics, geometry, and the concept of measure. Once again, we can see that the civilizing skill was interlinked with cosmology. In the ancient Chinese philosophy of Daoism we find the ordinal numbers represented through the act of counting, explicitly associated with the progressive stages of creation. We also see evidence of that same symbolic relationship in other ancient traditions. Even in the context of modern religion we see that associations between numbers and concepts of creation are retained, most evidently in the otherwise somewhat enigmatic Christian theme that God is One. The pervasive view is that Unity is a quality that is associated with the nonmaterial universe, sometimes designated as the Primordial Source. The act of counting is a sequential one and so implies that it must occur within the context of linear time, not an oscillating one. In fact, we could say that the act of counting becomes an effective metaphor for the progression of unidirectional, linear time.

One of the intriguing aspects of the ancient Egyptian hieroglyphic language is that the concept of nothingness is given by the word *tem/ temm,* phonetics that at the same time also reflects the cosmological concept of completion.[1] A related Dogon cosmological term, *toymu,* meaning "complete," is a counterpart to the ancient Egyptian word *temau,*[2] which means "all," "complete." The Dogon word arguably rests on the phoneme *te,* which means "only."[3] From this perspective, the primordial source of creation, represented as the nonmaterial universe, corresponds both to the numerical concept of zero (nothing) and simultaneously to the cosmological concepts of Unity and completion, in much the same way that the concept of infinity might. The same seemingly contradictory duality of outlook applies to concepts of primordial consciousness in the Tibetan Bon religion, whose cosmological terms consistently align with Dogon phonetics. There, consciousness is formulated as the Mind-Itself and is characterized all at once as being both empty (unknowable or ungraspable) and clear (intuitively comprehensible).[4] Similarly, the innermost cosmological knowledge of the Dogon, which is represented as ultimate, intuitive truth, is designated as the clear word.

We argued in *Seeking the Primordial,* based on comparative references from several traditions, that concepts of deity (from a Buddhist term *datu*) pertained to the aether unit—the one structure among the various component stages of creation defined by the Dogon that is not subject to change. Conceptually, the aether unit is the "egg" (or center point) of the Dogon egg-in-a-ball figure. This is the place that is designated as the domain of the creator god Amma and so is again associated with the concept of deity. As the first coherent structure of the progressive stages of material creation, it seems both mathematically sensible and affirming of the modern Christian view of deity to count the aether unit as "one." Dimensionally speaking, this center of spinning energy reflects the concept of a geometric point. If we take time, which we believe to be a dominant characteristic of the primordial source, as the first of the dimensional effects, and one that both originates and persists outside of the bounds of the dimensionality of space, then this first ordinal stage could be said to be one-dimensional. The existence of a point would be reflective only of the oscillation of time. Meanwhile, the oscillation of a dipole persistently cycles between "nothing" and "everything," much as the concept of primordial consciousness is said to do.

We argue that the aether unit is represented geometrically by the vesica piscis, the figure that is evoked by the alignment of a Buddhist stupa shrine. We also know that the shrine's geometry is derived using the cubit as a unit of measure, whose classic definition is given in relation to a person's forearm. A measured length is the first substantive, materially dimensional "thing." So it makes sense that an Egyptian term for *one* is given by Budge as *a,* meaning "piece," "one," "an,"[5] and symbolized by the bent arm glyph ⌐⎯⌐. We can take the Egyptian phoneme *a* as a counterpart to the Hebrew yah, which implies the concept of light—most especially since a ray of light would be the quintessential example of a two-dimensional measure (time and length). This concept is one that is also mimicked by a person's forearm—a line determined by two points. An Egyptian term for the concept of measure (and for the cubit itself as a unit of measure) was *meh.* The same term became a determinative for the concept of an ordinal number, signified by a glyph that again repeats the crisscrossing dynamic of angular impulse ∞. The shape also reflects how electromagnetism wraps when it intersects a flat surface, and how the Dogon specifically illustrate energy as wrapping at the "seat" of matter.

The concept of the number two is given in Egyptian hieroglyphics in relation to the nonmaterial and material universes. The term for the number two is *sen* (defined as "dual"),[6] a word that for both the Egyptians and the Dogon also implies the concept of a reflected image. We have said that for the Dogon, material reality represents a kind of reflected image. Sen is also the phonetic root of the Dogon term *sene,* which implies the cosmological concept of an electron or proton. Of course, the persistent molecule that is comprised of two electrons is a pair of hydrogen atoms and is one of our conceptual models for the Dogon term *Nummo*. The Egyptian word *sen* can also mean "counterpart," "companion," "neighbor," "colleague,"[7] terms that characterize the relationship between the two universes. Another Egyptian word, *sen,* means "thief" and so corresponds to the aspect of an electron in which it is "stolen" by another atom to form a molecular bond. In this same context, the term *sen* can also mean "to bind."[8]

We can argue that the Egyptian number three, which is given as *khemt,*[9] reflects the dynamics of instructed knowledge, both between the two universes and for the esoteric tradition. *Khemt* can also mean "to observe," "to think," "to think out a matter"[10] and "to not know," "ignorant."[11] These represent the obligations of an initiate and his or her informant in the esoteric tradition—to actively pursue an understanding of esoteric knowledge and to deny knowledge of inner secrets of the tradition if questioned by some uninitiated person. It is in this same context that we understand the Egyptian term *Khem* as an ancient name for Egypt itself. The same two effective meanings apply to the word "Dogon" and so identify both groups, through a shared naming convention, as the preservers of a common esoteric tradition. Khem is also the phonetic root of Egyptian words that can refer to a type of cloth and an unspecified part of a ship (perhaps a sail?) and so can be said to have conceptual ties to the notion of a square, which by our conception is the quintessential three-dimensional shape.

The Egyptian number four is expressed as *aft.*[12] Symbolically the word reads, "that to which ⌡ transmission ⟿ is given ⟿." The word is formed on the phonetic root *af,* which means "to turn," "to twist," "to revolve."[13] Symbolically the word reads "that which is ⌡ transmitted ⟿ as angular impulse ∝," and defines a glyph that is the image of a standing figure who holds a clay pot in its outstretched arm ⳨. Symbolically, this glyph illustrates the method by which the Dogon god Amma creates the world, and who is said to have raised and lowered his arm while turning in place to produce a helical shape. The clay pot (filled with water) would likely be the symbolic womb of the archaic tradition, a potbelly. Four is the symbolic number of the material universes and is also likely to be reflective of four dimensions. We are explicitly told by the Dogon that four is the number of the material universe. Similarly, the Egyptian word *aft* can refer to a rectangular box or chest, plot of ground, or rectangular stone,[14] configurations that are all, in our terms, four-dimensional.

We may not be capable of conceiving of the next higher dimension beyond the four material ones we experience, but we do have clues about some of its attributes—we can infer them from a kind of ordinal progression. First, it appears that each upward dimension adopts the shape of the prior dimension as its characteristic component. For example, each "slice" or conceptual segment of a line is a point; each component segment of a square is a line; each segment of a cube is a square. So we can infer that the next upward dimension from ours will somehow conceptually take a cube as its component segment, or its "slice." We also know that each of the three dimensions beyond a geometric point can be defined by a line, indicative of a spatial direction that sits perpendicularly to the previous dimension. We see that the ancient cosmology represents our four-dimensional reality with the figure of a square □, which combines four lines. In keeping with this idea, each of the Egyptian ordinal numbers can be represented symbolically by the appropriate number of vertical lines and so can be understood to reflect dimensional symbolism. (Likewise the concept of the nonmaterial, which we take to be one-dimensional, can be represented symbolically as a circle ○, a figure that consists of only one line without beginning or end. In that context, we might sensibly associate the circle with the concept of potential. So our reasonable expectation should be that the notion of a fifth dimension would define itself symbolically in relation to a figure with five lines. It is in this mode of thinking that we understand the ancient Egyptian word for the number five, which Budge defines as tu,[15] which can be alternately represented by the figure of a five-rayed star ⋆. The figure is in keeping with Dogon symbolism that interprets the spiral of matter in relation to seven rays of a star. The Egyptian word *tu* also refers to the concept of giving or givers.

When the Dogon speak of the seven pairs of universes, there is an implication that not all seven pairs necessarily exist concurrently with each other—in other words, that the formation of universes may themselves constitute a progression. The same inference would be in keeping with their description of the way in which Amma creates the universes, utilizing an up-and-down motion of his arm as he turns. From this perspective, we could see the creational dynamic of the universes as a correlate to angular impulse, which emerges as a progression and evokes a similar set of shapes. The same outlook could be in keeping with concepts in various ancient traditions of progressive ages.

An Egyptian icon for the number six is a clay pot ○. Egyptian and Dogon concepts of the number six play out in terms of phonetic roots and symbols that imply the notion of collecting, mixing, and sharing, functions that reasonably pertain to a clay pot. The idea of sharing seems like an appropriate counterpoint to the notion of giving that takes expression in relation to the number five. The Dogon term for six is *kuloy*.[16] It phonetically resembles a word that refers to the short sounding of a horn that serves as a signal for a group to come together. The same root is at the heart of a word for "cloud." What all of this might imply dimensionally may be beyond any sense of metaphor that would be clear to our four-dimensional viewpoint.

An Egyptian word for seven is *skhef*.[17] Symbolically it reads "translation ⃒ of the source ⊜ transmitted ⌔." At a microcosmic level, where the dynamic across a nonmaterial-material boundary seems to play out at a 1:7 ratio, the seventh effect completes the translated series. If, as the Dogon conceive, the seventh effect is the culmination of a word, then another Egyptian word, *skhef*, meaning "to write," makes sense.[18] What this dynamic may imply for universes is not entirely clear; however, the fact that the Dogon represent universes as a grouping of seven pairs indicates that it likely does carry implications on all upward scales of creation.

17

EXTENDED SYMBOLISM
OF LANGUAGE

BY THIS POINT IN our discussion, we should understand that the symbolic principles that define the concepts of the esoteric tradition also apply, somewhat unexpectedly, to much broader structures of the universe. As an example, dimensions seem to be treated as conceptual correlates both to DNA and to universes. The collapse of a primordial wave seems to relate to the dynamics of a black hole. Kabbalist references show that a Torah scroll is symbolic of the dynamic by which energy scrolls between the twin universes. We have seen how everyday concepts that pertain to ordinal numbers and the act of counting seem to apply symbolically to the dynamic of dimensions and universes. We know that ancient everyday units of physical measure can be said to symbolize tangible Earth measures, and ancient units of time appear to be factors of the grand cycle of precession. Similarly, we understand that sundials are reflective of the concept of a material arrow of time, while the notion of constellations may reflect how time is experienced nonmaterially. In other words, the principles of symbolism seem to apply to upward constructs of the esoteric tradition in the same way that the dynamics of matter can be demonstrated to recur on each of the upward scales of creation. Ultimately, symbolism seems to pervade all aspects of the esoteric tradition. So it should not surprise us to learn that many of these same effects also seem to apply to the structures of spoken and written language.

To the extent that our first evidence of the instructed cosmology is seen at Gobekli Tepe, symbolism seems to have been the original mode of expression of the tradition. The site is typified by carved images of animals and enigmatic carved symbols that also held cosmological significance for later cultures. In the mind-set of the cosmology, symbolism is treated as a natural and preferred mode for communicating ideas, if not perhaps also a necessary mode. Its demonstrated ability to correctly transmit information from generation to generation makes that an understandable viewpoint. We have said that the method of this symbolism is rooted firmly in our innate sense of parallelism—the understanding that two objects, actions, concepts, or processes are in some way fundamentally similar to one another.

From that perspective, any act of symbolism can be seen as an attempt to recast some concept that may be unknown to a person in terms that are likely to be more familiar to that person. In essence, if I know that you already have a good understanding of one of two parallel concepts, I can make use of that knowledge to explain the second concept to you. For example, a person who has no clear sense of what a black hole is might immediately grasp the essentials of its dynamic if I simply say that it behaves like a swirl of water as it goes down a drain. Looked at in this way, truly effective teaching in a symbolic mode consists largely of choosing the most apt metaphor to represent an idea. By that same token, effective interpretation of a symbol rests on the receiver's ability to intuitively grasp what the metaphor is about. To the extent that a person is able to do so, this quality of parallelism as an instructional tool is an inherently efficient one, in that it allows an instructor to teach a set of concepts once and then apply it many times. It also enhances the student's ability to learn when the nature of the instructional metaphors is consistent. Ultimately, the focus of the symbolic cosmology as an instructed tradition was on finding appropriate elements, drawn from an initiate's daily environment, to put forward as symbols, then using those to illustrate concepts of creation. The consistently inspired nature of these ancient symbols should demonstrate to us the innate capability of those who must have framed the cosmology. It also gives further testimony to the symbolic system as an instructed tradition.

The dynamics of language themselves constitute one of the symbolic elements of the esoteric tradition, in the context of a metaphoric theme that relates the processes of creation to the very act of speaking a word. This is the same familiar theme that we see in the Book of Genesis, where God merely speaks a phrase and by so doing creates light. Implicit in the theme is the scientific understanding that energy expresses itself through vibration, comparable to sound. A similar theme is represented in ancient Egypt with the idea that control could ostensibly be attained over a mythical entity simply by knowing its proper name. The same is true in a more practical sense for Dogon language in that the underlying meaning of a word is arguably reflected in its phonetics—more so for Egyptian hieroglyphic words, where specific nuances of meaning for a word seem to have been conveyed by the glyphs that were used to write it.

To the extent that the dynamics of creational energy express themselves through resonance, concepts of sound and music lend themselves as instructive metaphors. The seven notes of the Western musical scale are to sound what the seven colors of the rainbow are to light. Each is expressed as the consequence of a progressive series of vibrations. Similarly, the dynamic of speech or language becomes a familiar and intuitive metaphor for the expression of sound. Infants, as they grow older, learn to enunciate certain predictable, easily vocalized phonetic sounds as part of their process of acquiring language—*da, ba, pa, ma, ta,* and so on. These root sounds are very much of a kind with the root phonetics of the Dogon language. We could say that spoken language is to written language what walking is to running, in that mastery of one is effectively prerequisite to the other. In a normal developmental sequence, a person learns to speak before he or she learns to write. So it seems sensible that any progression of acquired civilizing skills would take its first expression as an oral tradition, not a written one.

Similarly, if we put ourselves in the place of an instructor of cosmology in some archaic, nontechnical age, we quickly understand that a lack of material resources must have been a factor that influenced certain instructional choices. As an example, Buddhist sources suggest that the alignment geometry of a stupa shrine was one of the first skills taught to a new group of initiates. As previously mentioned, Adrian Snodgrass states that the purpose of this exercise was to evoke an ordered space, also conceptualized as a sacred space, from the chaos of a disordered field, characterized as a profane space.[1] However, any capable teacher must have known that no tools would be readily at hand for the student/initiates to use to measure out this geometry. A likely solution to this difficulty was to define a unit of measure that centered on a student's own body parts, thus automatically guaranteeing, without regard to the availability of resources, that each student could participate. As mentioned previously, a cubit came to be defined either as the length of an initiate's forearm or as the span of their average pace or step, so that even those who might not be capable of taking measurements with their forearm might still participate.

Whereas the modern mind is preconditioned to work with units of measure that have been standardized to a precise length, the length of a cubit constituted a value that clearly varied from student to student. And so we understand that the cubit defines a relative unit of measure, not a precise one. However, we also understand that a variable unit of measure was appropriate to the task at hand, since the alignment geometry works its effect based on the relationship between the geometric figures, not as any consequence of their specific size. The choice of relative measurement makes sense, given that the dynamic of dimensionality itself is seen to be a by-product of a slowing time frame and so would therefore be of a variable nature. Again we could say that in this way the concept of a cubit also symbolizes a significant aspect of a universal dynamic. The ancient instructors of the symbolic tradition would have faced a similar supply difficulty with regard to cosmological language, in that there would have been no writing implements at hand or media available on which to even set words down in written form. And so it makes sense, simply from a standpoint of practicality, that the instructed civilizing plan in its earliest form would have been formulated as an oral tradition, not a written one.

In keeping with the societal imperative to preserve original forms, the Dogon language, which is also only spoken, not written, retains a very coherent set of cosmologically defined phonetic elements. An English-language speaker sees these largely as two- or three-letter syllables, each of which expresses what I call a *root concept*. Many of these same phonetic elements are commonly reflected in ancient Egyptian words, in the cosmological terminology of the Tibetan Bon tradition, and in the written Dongba language of the Na-Khi (or Na-Xi), from the borderland between Tibet and China. We often see these same phonetic elements reflected in ancient Chinese cosmological terms. We find many of the roots less obviously formulated, often in more complex ways, in Sanskrit. We see them consistently reflected in the names of Hindu deities. We find them in Hebrew cosmological terms. We see them in words of the Faroese language in northern Scotland. We often find them well preserved in the Polynesian language of the Maori in New Zealand. To one extent or another, we see evidence of them wherever we encounter other signature elements of the ancient cosmological tradition. We can think of these root syllables as conceptual building blocks that can be combined as compound words to express broader concepts. We often arrive at a better understanding of one of these compound Dogon terms simply by exploring the root meanings of the syllables that comprise it.

Cosmologically speaking, cross-cultural word comparisons are made with certain caveats. For instance, linguists understand that phonetics may evolve in certain predictable ways over time. There may also be specific phonetic values that are not reflected in the language of a given culture, or that perhaps do appear, but in an alternate form. For example, there are cases where the sound of an English-language *l* may seemingly be transposed to that of an *r*, that of a *p* may be expressed as an *f*, and so on. We have shown that words given by the Dogon and Egyptians with an *au* sound are arguably expressed in the Maori language as *whar*. Moreover, we have noted that Kabbalist sources recognize seven double letters of the Hebrew language whose pronunciations can shift under specific circumstances, comparable to the letter *pey/fey*. If we imagine these dual pronunciations to have been an original feature of the cosmology, then it seems possible that words formulated on these phonemes may have survived in later languages under just one of the two forms. We see this most obviously in words for "elephant," which may be given alternately as *pil* or *fil*, depending on which language we reference. Moreover, some of the earliest written languages, such as the Egyptian hieroglyphs or the Hebrew language, omitted written vowels entirely. This meant that those sounds were only inferred by the reader, not expressly written down—a feature that made written language only fully accessible to native speakers of the language, who would have already understood which vowel sounds to infer. Consequently it is not always certain to modern linguists how such ancient words may have been pronounced.

It is also often the case that modern researchers work from transliterations of words whose phonetics also may not always be consistent, and where the spellings can vary from source to source. For any researcher who attempts to correlate and trace original cosmological word forms, each of these variant spellings offers potential clues that can be of value to other researchers and so should be studiously preserved. Whereas in some fields of study standardization of terminology is taken as a worthy goal, for the comparative cosmologist the value lies in retaining each of the variant forms of words as they were represented in the earliest sources, even when those variations might tend to confuse a modern reader. The net of all of this is that, when making likely correlations between ancient words, we learn to allow a certain amount of wiggle room for reasonable differences—in other words, in the study of comparative cosmology we essentially have to learn to squint with our hearing.

With early symbolic written language, the choice to omit written vowel sounds seems like a counterintuitive one and so suggests that the practice could have had a symbolic motivation. Surely, written vowel sounds make a written word easier to read, and the choice to omit them creates an obstacle to reading. So from a practical perspective, the suggestion is of an attempt to restrict the ability of a text to be easily read. At the most obvious level of symbolic interpretation, the form of the language mimics one of the essential themes of the cosmology, in that in the mind-set of that cosmology, there are aspects to everyday reality that we are unable to readily perceive from a material view. Moreover, the very act of reimagining the vowel sounds while reading words might be seen as a practical exercise in inference (discriminating knowledge), one of the skills that the Dogon expressly say was to be fostered in humanity by the civilizing plan.

There is reason to think that the clustered meanings of ancient cosmological terms, which play out as homonyms in the Egyptian hieroglyphic language, may have also been utilized by ancient priests and scribes as a way to disguise the true meanings of certain esoteric passages, essentially as a kind of coding technique. In English this would be like representing the phrase, "I see the blue sky" in writing as, "Eye sea the blew ska." Read aloud, the translated passage could still reproduce understandable phonetics and convey correct meaning to someone knowledgeable in the tradition. However, to a person who might be unfamiliar with the spoken language or esoteric teachings, a literal translation of the passage would read nonsensically. This may explain why so many translators of Egyptian hieroglyphic texts end up choosing to interpret ancient passages somewhat poetically, perhaps in an effort to render what can seem nonsensical in a more sensible way. Our perspective is that proper translation may lie with an awareness of the clustered homonyms. Such coding techniques may have been seen as a way to preserve the integrity of secrecy for the esoteric tradition in what was seen as the inherently insecure mode of written media.

Quite often, the Dogon root syllables reflect the same clustering effect that we see with ancient Egyptian word meanings, as illustrated through multiple definitions in the *Dictionnaire Dogon*. We understand this effect to be either reflective or symbolic of the translational dynamic that defines the boundary between the nonmaterial and material domains. Because the Dogon language was documented in the 1950s as a living, spoken language, it provides us with a base of comparison for the often uncertain pronunciations of corresponding ancient Egyptian words, notwithstanding the normal permutations of a language over time. As an example, we equate the name of the Dogon Sigi/Sigui festival to an ancient Egyptian word, *skhai*, which means "to celebrate a festival." Similarly we correlate a Dogon word, *toymu*, meaning "complete," with the ancient Egyptian word *temau*. In many cases the word comparisons are more exact; the Dogon word *sene* associates with the Egyptian word *sen*, the word *po* aligns with the Egyptian word *pau*. Associations between such words allow us to gauge to what extent the Dogon pronunciations may have migrated during the intervening centuries and also help to establish an effective range of reasonable pronunciation for each original word. In any case we have seen that Budge's view of Egyptian hieroglyphic words consistently reflects the familiar set of Dogon phonetics, along with the clustered meanings that often attend them.

18

DISCRIMINATING KNOWLEDGE

WE HAVE SAID THAT one of the goals of the ancient civilizing plan was to foster a facility in humanity for discriminating knowledge. Most simply, this implies an ability to think for ourselves, to draw correct inferences from a given set of facts—in essence, the ability to make a logical deduction. Of course those skills require us to have established a degree of confidence in ourselves, in the pertinence of our own viewpoint, and in the quality and correctness of knowledge conveyed by the esoteric tradition. For these reasons, the idea of discriminating knowledge runs contrary to a popular notion that the attainment of higher spirituality requires people to somehow surrender their own sense of self, their personal discretion, or otherwise let go of ego. In the mind-set of the ancient cosmology, questions of ego more properly pertain to a discussion of material consciousness and how that relates to the idea of a primordial consciousness. These topics are perhaps given best expression in the Tibetan Bon religion, where primordial consciousness, referred to as the Mind-Itself, in its undifferentiated state, is said to defy the normal modes of material comprehension. The mind itself is represented to be, all at once, both *empty* (devoid of attributes) and *clear* (intuitively perceptible).[1] The effect that this produces is somewhat akin to the idea of an optical illusion, where two sets of visual elements of a figure compete for our active attention. One such famous image is that of the two mirrored faces, whose features evoke, in reverse image, the shape of a vase. The nature of human consciousness is such that when we focus with clarity on the two faces, the secondary image of the vase automatically fades from our attention, and vice-versa. Our mode of consciousness limits us to focused concentration on only one thing at a time and so effectively prevents us from actively experiencing both aspects of the image at once. The Bon tradition attests that, for us to arrive at a true sense of the Mind-Itself, we need to step out of learned modes of perception, which prioritize direct focus, and instead simply experience it, not actively work to deconstruct it. The dictate is one of perception, not of ego.

To the extent that the Bon tradition reflects the same cosmological symbolism that we have been pursuing, we gain insight into additional nuances of the key words *empty* and *clear* by exploring how the Dogon and ancient Egyptians understood the two terms. From one perspective, we could say that the words reflect the two extremes in oscillation of a dipole—one where polarized energy has expanded to its full extent ("empty") and the other where the energies come back together again ("clear"). In this context, an Egyptian term for "empty" can also mean "wide" or "spread out"[2] and so implies the concept of space or distance. The term rests on the phonetic root *us,* which also means "empty." Conceptually, the glyphs of the word read "the spiral ⌇ that binds ⟝∞⟞ distance ⛫ reverses △."[3] We see that the word is written with a glyph of two left-facing legs, which, from our perspective, implies the turning back of scrolling energy as it reaches its extended limit. Similarly, an Egyptian word for "clear," which Budge gives as *stef,* can also mean "to turn aside or away."[4] It reads "the dynamic ⧘ of matter ◠ is transmitted ⌇ forward △." Here we see a counter-reversal of energy implied by the use of the right-facing legs glyph. A similar dynamic of out-then-in oscillation is also conveyed by the Dogon word *vide,* which means "clear." The word can refer to "an abrupt waving of the hands as if to dismiss someone or something."[5]

We can say that two primary tools of discriminating knowledge are comparison and parallelism, concepts that are also first among the resources of a comparative cosmologist such as me. In keeping with a goal of fostering this skill of discrimination, we can interpret many of the symbolic structures of the cosmology as practical exercises in comparison and parallelism. For example, we have said that the very idea of a metaphor as a vehicle of expression, one that allows us to quickly grasp the meaning of one thing based on another, rests on understanding that the two referenced things are in some way parallel. Similarly the concept of duality, which is framed as a fundamental principle of creation, relies on a comparison of opposites. This concept allows us to intuitively comprehend a thing not through its similarity to some other familiar thing, but rather in opposition to it. The esoteric tradition's requirement that an initiate formulate the questions that advance his or her own initiation can be seen as an ongoing exercise in discriminating knowledge. At each stage of the instructed process students must essentially take stock of what they have already learned, then follow their own natural processes of synthesis to arrive at the next pertinent question. The initiate's facility for doing that is progressively honed by feedback from the instructor, who we have said simply defers from making a response until the student actually succeeds in framing a proper follow-on question. Through this back-and-forth dynamic, which is itself reminiscent of an oscillation, the student gains experience in and skill at framing discriminating questions.

At root, discriminating knowledge is a facility by which a person is able to utilize a set of facts to move beyond simple observation in order to make a conceptual leap. Such a leap in comprehension is often a natural outgrowth of mastery of a concept and ultimately qualifies a person to make certain judgments about the material itself. An excellent example of this kind of leap is found with the comparative architecture of original Neolithic stone houses at the Skara Brae village on

Orkney and stone houses as the Dogon still occasionally build them. Here the gathering of facts began with the simple observation that the two architectural forms are of the same kind. Due to its early origin, we know very little about the Orkney architecture beyond the mere fact and form of it, and an understanding that a single plan was applied to each of the original houses as a kind of archetype. However, for the Dogon, the house plan itself is understood to be a symbolic form. Through the configuration of its rooms, the house represents the body of a sleeping woman or goddess. This is in keeping with an important cosmological theme previously discussed in which the nonmaterial universe is characterized as a sleeping goddess and the processes of material creation are described as an awakening. Our conceptual leap came with recognition that the match in architecture implied that Orkney may have been influenced by the same symbolic cosmology as the Dogon, and in an era (ca. 3200 BCE) that just precedes the one we infer, based on comparisons to ancient Egypt, for the Dogon societal practices (ca. 3000 BCE). This realization opened a door that allowed us to compare evidence on Orkney to other aspects of the Dogon cosmology. It also justified comparing ancient words and place-names on Orkney to pertinent Dogon and Egyptian words. Given that, it also made sense to examine references in relation to the two ancient languages associated with Orkney, which were the Norn and Faroese languages. In essence, the objective similarities of two architectural forms facilitated, through deductive inference, an entire study.

A second example of this kind of conceptual leap is seen in the previously mentioned ancient Egyptian and Chinese hieroglyphic words for "week." Because each word was written with only two glyphs, they are among the most accessible of word forms for a neophyte in ancient symbolic language to explore. With the benefit of perfect 20/20 hindsight, these words would have been our entry point to the language, in that it should be apparent to any clear-thinking symbolist that when we combine a symbol for the concept of a day ⊙ with the number 10 ∩ we convey a meaning of "ten days," which is the very definition of an ancient week in both cultures. Even more helpful to the interpretation of this word is the fact that the notion of a week is not a nuanced concept, but rather a discrete and objective one. One of the conceptual leaps that arises from an understanding of this word comes when we consider whether similar principles of glyph construction could also apply to other ancient hieroglyphic words; in our estimation, they do to virtually all ancient Egyptian words, as Budge presents them. Another comes when we grasp that this single word comparison intuitively demonstrates that the underlying principles that governed the ancient Egyptian and Chinese hieroglyphic languages must have been the same at the time of their formulation. In these ways, the word becomes an anchor point for the two traditions. Again, an entire study of symbolic language is potentially facilitated by the conceptual leap.

Much as good parents come to realize that their infant's ability to "catch on" to how things work in daily life improves when the parents make an effort to do things in consistent ways, so the framers of the esoteric tradition understood the value of consistency when attempting to convey conceptual information to some future generation. It is because of this that a broad familiarity with the mind-set of the tradition improves our ability to discern esoteric meaning. It is this sensibility that accounts for the consistency of symbolic forms that we have discussed. It is for this same reason that, beginning with the cosmological theme of, "As above, so below," we see consistent emphasis placed on the parallelism of cosmological themes and symbols. Once we catch on to this aspect of the tradition, we should know after only a few examples to look for the root meaning of symbolism in the salient attributes of the animal, object, or element that is featured. For instance, we know that water is said to be of the nature of the nonmaterial domain, and Earth is symbolic of the material domain. The Dogon classify a dung beetle among the other water beetles. This is a creature that moves from the water to the land and so sensibly comes to symbolize the concept of nonexistence coming into existence. A bird of prey grasps things with its talons and so comes to be associated with creator deities whose function is to grasp energy, hold it firm, and shape it. A serpent undulates as it slithers and so mimics the flow of energy. We make another conceptual leap when we learn to interpret which attribute of an animal pertains to its likely symbolism. We are often able to affirm our view of that by looking first to the root phonetics of the name of the animal in Dogon culture or to the glyphs that comprise its written name in the Egyptian hieroglyphic language.

One of the prerequisites to an understanding of the esoteric tradition rests with a person's ability to entertain the notion of a nonmaterial domain, one that is, in many ways, hidden from our view. As a student I was taught that a scientist's job was to objectively explore each of the prospective solutions to a question, not to prejudge them—especially when some of those potential solutions might go against one's own preconceptions. When faced with an unlikely implication, the job of an objective researcher is to consider from what perspectives that implication might possibly be true, rather than to make rote declarations of why it cannot possibly be. The first approach fosters a mind-set of open exploration of a subject, the other one of simple denial. I often say that the first thing a person in my field of study learns is that there are things going on in the world that we cannot yet explain. Likewise, one of the first conceptual leaps that is required of a researcher in this field involves a willingness to allow that things foreign to our everyday experience might have a degree of reality to them, even if they seem to run counter to traditional perspectives.

Consistent with the requirement for secrecy in the esoteric tradition, the Dogon employ certain techniques for discretely pointing us to essential knowledge, in much the same way that an informant in the esoteric tradition effectively guides a student to the next question. The Dogon cosmology often declares two processes or sequences to be parallel, outlines the details of the sequence for one of the processes, but then only goes on to discuss selected details of the comparative process, simply omitting other, apparently more secretive details. Such comparisons are ripe for our examination in that they invite us to simply infer the missing elements of the comparative sequence.

As an example, when we are told, in the context of a discussion of Dogon numerology, that three is the number of the male, four the number of the female, and seven the number of the individual, a pattern has been set to interpret (allowing for symbolic reversals) what is later implied but not actually stated about parallel numerology of the two universes. We are expressly told that four is the number of the material universe, an assignment that seems sensible since material existence takes its definition in relation to four dimensions. We also know that masculinity is associated with the material universe and femininity with the nonmaterial domain, so the unstated implication, dictated by parallelism, is that three must be the number of the nonmaterial universe. The conceptual leap that requires some discrimination on the part of the observer rests with inferring that since seven symbolically defines the number of the individual, the two universes together might therefore also constitute an individual. As it turns out, both the ancient Egyptians and the Kabbalists are in specific agreement with the outlook—they each document an explicit concept of a nondeified *primordial individual,* essentially a universal consciousness that is comparable to the Bon religion's notion of the Mind-Itself. Budge tentatively assigns the name of *sah* to this concept, along with a complex identifying figure that combines the Eye of Horus (symbolic of the nonmaterial universe) and the Eye of Ra (symbolic of the material universe), set above the figures of seven stars, set in a line. Two facing eyes, often termed Wadjat Eyes, were a notable feature that was often painted on the outside of the wooden coffins of pharaohs during the Middle Kingdom of Egypt, in the centuries following 2000 BCE (see fig. 18.1 on p. 174).

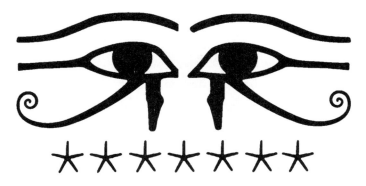

Fig. 18.1. Sah, the Spirit body of Orion
(Budge, p. 646a)

In this same context, we can cite another example of how discriminating knowledge fosters a more sensible interpretation of cosmological meaning. We understand, based on the root phonetics of the cosmology, that the syllable Si applies symbolically to the non-material universe and Ra, along with the shape of the sun glyph, to the material universe, so that a cosmologically appropriate rendering of the ancient Egyptian name, which Budge interprets as Sah, would likely be Si Ra or Sirah. The outlook complies with the mind-set of cosmological language, where root syllables are combined to define larger concepts. It also is in keeping with Schwaller's Islamic notion of the Bridge of Sirah, the idea that truth will be found to lie on a razor's edge between two chasms, symbolic of science and superstition. As it turns out, that razor's edge, akin to the thin surface of water, coincides with Samkhya and the Dogon's interface between the nonmaterial (Si) and material (Ra) domains.

If we take another giant step back from our cosmological view of things, we understand that the dynamics of discriminating knowledge are also the essential dynamics of consciousness itself. Our sense of how the brain establishes meaning rests on the same back-and-forth comparative dynamic between two perspectives of overview and detail. We can imagine that this handoff of perspectives continues until the meaning of a thing comes to be comprehensible. In fact, one of my personal pet theories has long been that humor is found in those things that temporarily confuse this back-and-forth dynamic. The idea is that when it is not immediately clear whether a thing should be better understood in overview or in detail a kind of negotiation occurs between the hemispheres of the brain. Laughter would be what happens during that period of negotiation.

FINAL THOUGHTS ON THE COSMOLOGICAL PLAN

EACH BOOK IN THIS series on comparative cosmology has added the weight of new evidence that implies the influence of an ancient instructed civilizing plan on the creation traditions of numerous widespread cultures. In this volume, our focus has been on aspects of the symbolic system that demonstrate intention, and so, in keeping with Dogon and Buddhist views of their symbolic cosmologies as having been instructed, they reflect a set of pedagogic choices that were seemingly made while formulating the ancient symbolic system. Each apparent choice offers us a degree of insight into the thought processes that produced it. As we work with the ancient cosmology, it becomes apparent that we are exploring a system that was thoughtfully set out, not just some haphazard assemblage of incidental elements. Rather we see indications of a scientifically based cosmology, integrated with a civilizing plan whose immediate practical focus was on an instructed agriculture.

References to the tradition might strike us at the outset as enigmatic but ultimately gain in sensibility as we become more familiar with them. If, in fact, we are dealing with an instructed tradition, then surely the various contributing elements take their most coherent form as they were originally introduced, not as they may survive among widespread cultures thousands of years later. For that reason, one of the keys to untangling the Gordian knot of ancient cosmology rests with finding effective ways to trace original forms, which is one of the goals of comparative study. To that purpose, the Dogon, whose societal emphasis happens to be on the preservation of original forms, seem like a very natural and sensible entry point from which to explore them. Researchers from many different fields of study, from British polymath Thomas Young and French scholar Jean-Francois Champollion, in their successive efforts to decipher Egyptian hieroglyphs in the nineteenth century, to Carl Jung with his study of ancient archetypes in the twentieth, have expressed their own determined sense of something deeper and more essential to this class of ancient symbols, themes, and myth than first meets the eye, a tradition that has long been shrouded in mystery. It has been this same abiding enigma, reflected in so many aspects of ancient studies, that has sustained our focus on the esoteric tradition and drawn us all to consider it more deeply.

In keeping with the idea of an instructed tradition, the mind-set of Dogon culture is a distinctly nonjudgmental one—of the sort that we might expect from a fair-minded educator, not of a dogmatic priest. There is no emphasis in Dogon culture on familiar themes of some modern religions—no notion of Good versus Evil—and therefore no working concept of a heaven or a hell, nor of redemption or damnation. Interestingly, despite other parallels to cosmological traditions in India, the Dogon specifically have no sense of reincarnation, at least not any that goes beyond the genetic passing of a person's traits to offspring. In keeping with the nonjudgmental outlook of Dogon cosmology, we learn that there is no consequence for an initiate to the esoteric tradition whose line of questioning misses the expected mark of his or her informant, beyond sincere encouragement to simply try again. Similarly, the usual consequence for a Dogon tribesperson who may become involved in a village dispute is an immediate opportunity to resolve the concern, actively facilitated by the community itself. In keeping with this level outlook, a typical Dogon village is a notably peaceful place. All of this is in keeping with what we might expect from an archaic matriarchal authority that seems, by virtually every account, to have taken a sincere maternal interest in the long-term development of humanity.

Perhaps our first clue to deliberate intention within the symbolic system comes with the observation of consistency of method. We see it most obviously in its archaic form as an oral tradition, evident in the way that terms of the cosmology are formulated from root syllables whose meanings hold true from word to word. This is a circumstance that we might not expect to see with a language that grew up piecemeal over time, and we surely should not expect to find it just incidentally arise across numerous unrelated ancient cultures. Moreover, meanings associate with phonetics and the formulation of spoken words in the same way that they do with early symbolic written characters of the Chinese hieroglyphic language, and by the very same method of formulation that we observe for the Egyptian hieroglyphs. Beyond that, if we endorse the concept of an Egyptian *defining word* (words that seem to assign symbolic meanings to individual glyphs), then intention as reflected by cosmological language becomes a very difficult thing to deny.

We observe a number of fairly sophisticated techniques of instruction consistently evidenced in the archaic cosmology. From the outset, intention is apparent in the fireside myths of Dogon cosmology, whose audience consists of as yet uninitiated tribespeople, and whose plotlines relate to themes of cosmology, not to those of morality plays, fables, or soap operas. When we pursue concepts, facts, relationships, and dynamics that are relayed to us in these myths, we find that they lead us to credible information about concepts of creation that are scientifically pertinent. When the focus of a myth is on an energetic creational process, it becomes hard not to infer that the role of the myth is instructional—again realizing that any act of deliberate instruction reflects intention. Intention is similarly reflected in many of the practices that define daily life for the Dogon, many of which Griaule and Dieterlen also relate to cosmological processes. As an example, noted above, the archaic layout of a Dogon agricultural plot was said to recreate the dynamics of the spiral of matter. In later eras, and in keeping with other symbolic reversals of the tradition, the Dogon agricultural plots took on a squared configuration, in accordance with a specific cosmologically based scheme known as the well-field plan, one that has also been documented for ancient China.

The techniques by which symbolism is conveyed in the Dogon esoteric tradition also show clear intention in their formulation. From the broadest view, cosmological concepts are discussed in the context of three parallel creational themes: how the universe forms, how matter forms, and how the processes of biological reproduction happen. In what can only be described as a stunning demonstration of both intention and capability, the cosmology simultaneously represents all three themes through a single progression of symbols. The choice to frame creational concepts in this way—as parallel sequences—sets the stage for the use of parallelism as an ongoing instructional device throughout the entire symbolic cosmology. The choice of parallelism as an instructional tool is an efficient one in that it makes it possible to define a given concept once and then apply that definition many times. Meanwhile, the use of a variety of well-conceived metaphors as a technique to convey knowledge takes its instructional advantage from this very same dynamic of parallelism. The various four-stage metaphors through which the symbolism is expressed are themselves framed in parallel with each other and so also reflect intention and consistency of method on the part of those who formulated them.

In this field of study, interpretations often begin with outward resemblances, and the work of a researcher is often simply to try to anchor those resemblances with specific ancient references. Until and unless some ancient culture can be shown to have interpreted things the way we are inclined to see them, all we truly accomplish by promoting a resemblance is the chasing of unicorns. However, the structure of the cosmology provides us with a basis for framing certain arguments based on the parallel nature of structures and processes. In the end we come to understand that parallelism of the dynamics implies that we will see ongoing resemblances on each of the upward scales of creation, and simply in that context, those resemblances can be seen as legitimate. Our observation is that the symbolic tradition had the perspective to take note of those parallels, even so far as to frame them as a Hermetic theme, and the ancients also often took specific notice of and commented on them.

Just as parallelism is a shortcut to mastering a new concept, so is our understanding of the principle of opposites. Just as a person can intuitively grasp the situation when thing A is very like thing B, so we can arrive at immediate comprehension if thing B is represented as being opposite to thing A. Such comparisons take advantage of our preexisting knowledge to directly convey ideas that might otherwise be difficult to represent. As an example, I can effectively impart the concept of stillness to a person simply by framing it in opposition to a state of motion. Emphasis within the tradition on the principles of duality and the pairing of opposites similarly orient us to actively look for relationships between things that are either parallel or opposite. Moreover, since in the mind of Samkhya these same principles of parallelism and opposition pertain to the state of the paired universes, they again become symbolic of creational dynamics on a broader scale.

Similarly, fluency with the concept of a progression provides yet another effective tool for the efficient communication of ideas. Perhaps the most intuitive example of a progression is illustrated by the simple act of counting, and so the familiar sequence of ordinal numbers becomes a frequent mode of symbolism in many ancient traditions. We also see that much of the symbolism of the cosmology corresponds to progressions in nature that are familiar to us—for example, how a plant grows from a seed, or how an egg comes to produce an adult goose. These are metaphors that would also be particularly familiar and useful in the context of instructed agriculture or animal husbandry. Again, the underlying effect rests on parallelism between the stages of a cosmological process and those of an everyday progression. The idea is that the symbolically equated progression will effectively guide our understanding of the cosmological process in a more intuitive way.

Another significant instructional dynamic of the ancient symbolic system takes its benefit from the concept of a continuum—the idea that a given process can be understood to be both linear and progressive. The underlying concept is basically that of a bus route and our understanding that if a passenger gets on the bus *here* and is never seen to disembark, that passenger must eventually arrive *there*. Often it is the concept of a continuum that ultimately makes an inference possible, which is why inferences constitute one of the powerful end expressions of discriminating thought. Cosmologically speaking, the relationship between increased mass and slowing time frame (Einstein's concept of relativity) is the quintessential scientific example of a continuum—an ongoing process which, unless demonstrably interrupted in some way, points us to a specific outcome. It is the dynamic of the continuum of relativity that insists that any nonmaterial domain must experience an ultimate quickness of time frame compared to ours, a circumstance that carries specific implications. Acknowledgment of this necessary aspect of nonmateriality points us to that same effect as a potential explanation for other elusive effects, like those of quantum entanglement. Ancient themes associated with the nonmaterial, such as Unity, also become sensible in the context of an ultra-quick time frame.

We know that spinning energy (angular momentum) is the dynamic that imparts mass and is responsible for slowing time frame. Meanwhile, the associated effect of angular inertia provides us with a credible catalyst for the cosmological dynamic of Unity becoming Multiplicity. Not only do these ancient principles of creation fall into a reasonable scientific context, they relate to a very understandable and manageable set of effects of energy that all cluster together, conceptually speaking. Moreover, they align quite precisely with Dogon descriptions and drawings that ostensibly reflect root dynamics of creation.

The concept of rituals that recur, along with the institution of holiday celebrations, promotes an understanding of the concept of a cycle, which again represents one of the underlying dynamics of creation. Recognition of the attributes and effects of a cycle can also enhance a person's understanding of how a process works. The oscillation of a dipole is perhaps the quintessential example of a cycle and perhaps represents the natural underlying state of energy. The same notion is promoted through cosmological emphasis on the natural water cycle, which is offered by the Dogon and others as a correlate to the cyclical flow of energy that scrolls between the universes. Similarly, the act of ritual sacrifice is specifically represented by the Dogon through the idea that blood spilled from a slaughtered animal symbolically serves to complete an energetic cycle, so again directs our attention to an important creational dynamic. Beyond that, specific celebrations, such as the fifty-year Dogon festival of the Sirius stars, point us to important effects of angular momentum as underlying causes for those dynamics.

Each of these instructional tools helps to promote a facility for discriminating knowledge, which the Dogon flatly say was one of the underlying purposes of the civilizing plan. The dynamic of knowing, itself, is ultimately arrived at through acts of comparison, and start (as my wife Risa often points out) with the simple question of whether some newly discovered thing is like or unlike some other thing a person has previously encountered. Ancient symbolism, as a more directed mode of expression, rests on the idea that the symbol is somehow intimately like the thing it represents (the Dogon say that the symbol is actually more essential than the thing itself), and so one of the ongoing challenges for those who interpret symbols rests with sorting out how two symbolically correlated elements can be understood to be intimately alike. To the extent that the focus of the ancient symbolism was on creational processes of the universe, it seems to have been hoped that an outgrowth for humanity of knowledge of those processes would be a clearer perspective on what our own relationship is to those processes. The clear expectation was that, once aware of the broader picture, humanity could make better choices about what we do and how we conduct ourselves.

Exoteric (public) Buddhism, like many other familiar traditions, represents that the root dynamics of creation are inherently unknowable and so must remain incomprehensible to humanity. However, we have come to see such statements as among the occasional deliberate fibs of the ancient cosmological tradition. By contrast, esoteric Buddhist thought, which is well reflected in Shingon Buddhism, confesses that the root dynamics of material creation are entirely knowable. At heart, the structures of material creation appear to be a consequence of polarized energy, whose tendency it is to spin and oscillate. Oscillation is a persistent inward and outward motion that serves as the primordial metronome of time. Like a musical metronome, changes in mass give time the capacity to oscillate at various rates, and its frequency depends largely on the amount of resistance that the oscillating energy meets. Meanwhile, polarized energy also has the tendency to spin, and as it spins it evokes mass. It is the nature of mass to provide the resistance that slows the rate of time's oscillation.

In each of the above cases, it seems clear that someone took great care with how the civilizing plan was formulated and presented. What is indicated at each stage is the hand of a skilled pedagogue who not only demonstrated full mastery of a group of complex subjects, but also a great deal of insight into how best to foster learning in people. Moreover, the symbolic system also shows an awareness of humanity's tendency to resist new knowledge, alongside a reasonable set of psychological strategies aimed at offsetting some of that resistance. Whoever formulated the instructional plan seems to have understood that later generations might have difficulty validating some of the root concepts of creation, which meant that these aspects of the tradition would likely have to be entertained, at least initially, based on a sense of the authority of the teacher. And so care was taken with the more accessible material to establish a credential for correctness of perspective, largely by bringing symbolic concepts down to a core group of unmistakable shapes and descriptions that would lend themselves to side-by-side comparison with scientific counterparts. It is for this reason that Buddhist thought emphasizes the significance of adequate symbols, those whose meanings align with creational shapes that persist in nature. The mere inclusion of such shapes among the symbology of ancient cultures, described in proper scientific relationship to each other and in terms wholly appropriate to their creational themes, demonstrates again the very great capabilities of those who formulated and promoted the ancient cosmology. Many of these same shapes survive as cosmological symbols in Dogon culture and provide a coherent point of reference by which to understand Dogon discussions of their egg-in-a-ball and egg-of-the-world symbolic constructs. The Dogon concepts relate to effects of energy that operate within the same microcosmic domain as Buddhist adequate symbols, a realm that would remain inaccessible to observation by any society that hadn't already attained a certain degree of technological capability.

The compelling promise that the esoteric tradition extends to humanity is to provide us with the conceptual keys to many of the root processes of creation. But that goal would not really be attainable without first bringing us to an awareness of the underlying nature of the universe. Through the eyes of the archaic Samkhya cosmological philosophy we are told that universes form in pairs. Comparative studies imply that these paired universes function energetically much like two hemispheres of a brain and so are expressly treated in various ancient traditions as a nondeified, primordial, conscious individual. In the mind-set of the tradition, that consciousness is effectively a macrocosmic counterpart to a conscious human individual, much in the same way that we understand Barnard's Loop to be a macrocosmic counterpart to the tiny spiral of matter. In the view of Samkhya, it is the nature of this grand consciousness, like any other self-aware being, to call out for communication and companionship. The essential reality is that there is only one potential pool of candidate companions for a nonmaterial consciousness—namely, the individualized consciousnesses that are said to be evoked within its material twin universe. Moreover, at the depth of the scrolling cycle of energy, when quickness of time frame for the nonmaterial universe implies that all events may occur at once, and so it is left with no productive moment in which take coherent action, this primordial consciousness, like a fully aware but paralyzed person, would also seem to require caretaking. For this reason one of the implied motives of the esoteric tradition seems to be to select and train potential caretakers who are both aware of and sensitive to the ultimate plight of the nonmaterial consciousness.

If we entertain these goals as having been the primary ends of the esoteric tradition, then the various symbolic structures that comprise the tradition seem more than sensible. In a circumstance where speech in our familiar sense was not an option, symbolism was the language by which the underlying message was conveyed, and so the closely linked civilizing structures worked to preserve and promote an understanding of that mode of communication. The archaically instructed cosmological knowledge seems to have been targeted to the era in which we presently live—one that has achieved a modicum of its own technological understanding. This particular era also seems to be postured at a point of transition for the cycle of scrolling energy. In the context of the Yuga Cycle, we understand that without specific supporting societal structures a fully descended material universe might not even be capable of perceiving the presence of a nonmaterial domain, let alone interacting productively with it. The symbolic cosmology, paired with its companion set of civilizing structures, work together on behalf of all involved to preserve the memory of that now largely imperceptible universe. And because the cyclical nature of the relationship implies that each universe takes its alternate turn at descending and ascending, there is an ongoing mutual obligation for each to play the role either as informant or companion for the other, and thereby perpetuate what we see as the plan of the ancient cosmology.

NOTES

1. MOTIVES AND INTENTIONS OF THE ESOTERIC TRADITION

1. Saraswati, *Samkhya Darshan*, 62.
2. Budge, *An Egyptian Hieroglyphic Dictionary*, 178ab.

2. WHAT MAKES US THINK THERE WAS A PLAN?

1. Grimes, *Ganapati: Song of the Self*, xiv.

4. METAPHORS OF THE COSMOLOGY

1. Budge, *An Egyptian Hieroglyphic Dictionary*, 11ab.

5. THE ALIGNED SHRINE

1. Snodgrass, *The Symbolism of the Stupa*, 1.

7. THE ROLE OF MYTH

1. Snodgrass, *The Symbolism of the Stupa*, 6.

8. SYMBOLIC ASPECTS OF ANGULAR MOMENTUM

1. Snodgrass, *The Symbolism of the Stupa*, 252.
2. Budge, *An Egyptian Hieroglyphic Dictionary*, 469b.
3. Budge, *An Egyptian Hieroglyphic Dictionary*, 469a.

9. SYMBOLISM OF TIME AND SPACE

1. Snodgrass, *The Symbolism of the Stupa*, 39.

2. Snodgrass, *The Symbolism of the Stupa,* 62.

3. "Ganesha," article available on McGill University website. Search on "McGill Ganesha wiki."

4. Grewal, *Book of Ganesha,* 89.

5. Calame-Griaule, *Dictionnaire Dogon,* 297, and Budge, *An Egyptian Hieroglyphic Dictionary,* 420ab.

10. MYTHOLOGY OF LIGHT

1. Snodgrass, *The Symbolism of the Stupa,* 287.

2. Snodgrass, *The Symbolism of the Stupa,* 193.

3. Snodgrass, *The Symbolism of the Stupa,* 210.

4. Calame-Griaule, *Dictionnaire Dogon,* 303.

5. Calame-Griaule, *Dictionnaire Dogon,* 301.

6. Calame-Griaule, *Dictionnaire Dogon,* 303

7. Budge, *An Egyptian Hieroglyphic Dictionary,* 5a.

8. Where not otherwise noted, supporting material relating to Ogo is drawn from chapter 2, "Ogo," of Griaule and Dieterlen's *The Pale Fox.*

9. Budge, *An Egyptian Hieroglyphic Dictionary,* 23a.

10. Budge, *An Egyptian Hieroglyphic Dictionary,* 99a.

11. Calame-Griaule, *Dictionnaire Dogon,* 209–10.

12. Griaule and Dieterlen, *The Pale Fox,* 198.

13. Budge, *An Egyptian Hieroglyphic Dictionary,* lxviii and cxxxiii.

14. Budge, *An Egyptian Hieroglyphic Dictionary,* cxviii.

15. Griaule and Dieterlen, *The Pale Fox,* 198.

16. Scranton, *The Cosmological Origins of Myth and Symbol,* chapter 16.

17. Budge, *An Egyptian Hieroglyphic Dictionary,* 25a.

18. Calame-Griaule, *Dictionnaire Dogon,* 209–10.

11. LESSONS
IN SACRED GEOMETRY

1. Snodgrass, *The Symbolism of the Stupa,* 28.

2. Snodgrass, *The Symbolism of the Stupa,* 153.

12. NONMATERIAL
TO MATERIAL TRANSLATION

1. Budge, *An Egyptian Hieroglyphic Dictionary,* 104b.

2. Budge, *An Egyptian Hieroglyphic Dictionary,* 159a.

3. Budge, *An Egyptian Hieroglyphic Dictionary,* 522b.

4. Budge, *An Egyptian Hieroglyphic Dictionary,* 34b.

5. Budge, *An Egyptian Hieroglyphic Dictionary,* 419b.

6. Budge, *An Egyptian Hieroglyphic Dictionary,* 9b.

7. Budge, *An Egyptian Hieroglyphic Dictionary,* 96a.

8. Budge, *An Egyptian Hieroglyphic Dictionary,* 105a.

9. Budge, *An Egyptian Hieroglyphic Dictionary,* 665b.

13. SELF-CONFIRMATION
OF MEANING

1. Budge, *An Egyptian Hieroglyphic Dictionary,* 376b.

2. Scranton, *China's Cosmological Prehistory,* chapter 3.

3. Budge, *An Egyptian Hieroglyphic Dictionary,* 331a.

14. DYNAMIC OF THE
INITIATE AND INFORMANT

1. Budge, *An Egyptian Hieroglyphic Dictionary,* 712a.

2. Budge, *An Egyptian Hieroglyphic Dictionary,* 451b.

15. THE NATURE OF WATER

1. Budge, *An Egyptian Hieroglyphic Dictionary,* 841b.

2. Budge, *An Egyptian Hieroglyphic Dictionary,* 293a.

3. Budge, *An Egyptian Hieroglyphic Dictionary,* 400a.

4. "Why Is Water Considered a Dipole-Dipole Force?" question posted on the Quora website; answer provided by "Natasha Hesketh, Teacher" on April 30, 2018.

5. Budge, *An Egyptian Hieroglyphic Dictionary,* 349b.

6. Ohno, Fujita, and Khono, "Is Seven the Minimum Number of Water Molecules per Ion Pair for Assured Biological Activity in Ionic Liquid–Water Mixtures."

16. UNITY AND
THE DIMENSIONALITY OF NUMBERS

1. Budge, *An Egyptian Hieroglyphic Dictionary,* 834a and 835a.

2. Budge, *An Egyptian Hieroglyphic Dictionary,* 834a.

3. Calame-Griaule, *Dictionnaire Dogon,* 269.

4. Rossi, *The Philosophical View of the Great Perfection in the Tibetan Bon Religion,* 95.

5. Budge, *An Egyptian Hieroglyphic Dictionary,* 105a.

6. Budge, *An Egyptian Hieroglyphic Dictionary,* 673b.

7. Budge, *An Egyptian Hieroglyphic Dictionary,* 673b.

8. Budge, *An Egyptian Hieroglyphic Dictionary,* 675b.

9. Budge, *An Egyptian Hieroglyphic Dictionary,* 543b.

10. Budge, *An Egyptian Hieroglyphic Dictionary,* 543a.

11. Budge, *An Egyptian Hieroglyphic Dictionary,* 543a.

12. Budge, *An Egyptian Hieroglyphic Dictionary,* 44a.

13. Budge, *An Egyptian Hieroglyphic Dictionary,* 43b.

14. Budge, *An Egyptian Hieroglyphic Dictionary,* 43b.

15. Budge, *An Egyptian Hieroglyphic Dictionary,* 868b.

16. Calame-Griaule, *Dictionnaire Dogon,* 171.

17. Budge, *An Egyptian Hieroglyphic Dictionary,* 690a.

18. Budge, *An Egyptian Hieroglyphic Dictionary,* 690a.

17. EXTENDED
SYMBOLISM OF LANGUAGE

1. Snodgrass, *The Symbolism of the Stupa,* 14–15.

18. DISCRIMINATING KNOWLEDGE

1. Rossi, *The Philosophical View of the Great Perfection in the Tibetan Bon Religion,* 95.

2. Budge, *An Egyptian Hieroglyphic Dictionary,* 182b.

3. Budge, *An Egyptian Hieroglyphic Dictionary,* 181a.

4. Budge, *An Egyptian Hieroglyphic Dictionary,* 11a.

5. Calame-Griaule, *Dictionnaire Dogon,* 298.

BIBLIOGRAPHY

Brighenti, Francesco. *Sakti Cult in Orissa*. New Delhi: D. K. Printworld Pvt. Ltd., 1963.

Budge, E. A. Wallis. *An Egyptian Hieroglyphic Dictionary*. New York: Dover Publications, 1978.

Calame-Griaule, Geneviève. *Dictionnaire Dogon*. Paris: Librairie C. Klincksieck, 1968.

Clark, R. T. Rundle. *Myth and Symbolism in Ancient Egypt*. London: Thames and Hudson, 1959.

Dunrea, Olivier. *Skara Brae: The Story of a Prehistoric Village*. New York: Holiday House, 1985.

Forde, Daryll, ed. *African Worlds: Studies in the Cosmological Ideas and Social Values of African Peoples*. Oxford: Oxford University Press, 1954.

Grewal, Royina. *The Book of Ganesha*. New York: Penguin, 2003.

Griaule, Marcel. *Conversations with Ogotemmeli*. Oxford: Oxford University Press, 1970.

Griaule, Marcel, and Germaine Dieterlen. *The Pale Fox*. Paris: Continuum Foundation, 1986.

Grimes, John A. *Ganapati: Song of the Self*. Albany: State University of New York Press, 1995.,

Horowitz, Edward. *How the Hebrew Language Grew*. New York: KTAV Publishing House, Inc., 1960.

Kemp, Barry J. *Ancient Egypt: Anatomy of a Civilization*. London and New York: Routledge, 1989.

Ohno, Hiroyuki, Kyoko Fujita, and Yuki Khono. "Is Seven the Minimum Number of Water Molecules per Ion Pair for Assured Biological Activity in Ionic Liquid–Water Mixtures?" *Physical Chemistry Chemical Physics* 22 (2015).

Ouaknin, Marc-Alain. *Symbols of Judaism*. New York: Assouline, 2000.

Parker, Philip M. *Webster's Faroese-English Thesaurus Dictionary*. Las Vegas, Nev.: ICON Group International, Inc., 2008.

Reid, Marcus. "Virtual Particles in Electromagnetism." PDF posted on Vakuumenergie (German website).

Rice, Michael. *Egypt's Making: The Origins of Ancient Egypt 5000–2000 BC*. New York: Routledge, 1990.

Rossi, Donatella. *The Philosophical View of the Great Perfection in the Tibetan Bon Religion*. Boston and London: Snow Lion, 1999.

Saraswati, Swami Niranjanananda. *Samkhya Darshan: Yogic Perspective on Theories of Realism*. Munger, Bilar, India: Yoga Publications Trust, 2008.

Sauneron, Serge. *The Priests of Ancient Egypt*. Rev ed. Ithaca, N.Y.: Cornell University Press, 2000.

Scholem, Gershom. *On the Kabbalah and Its Symbolism*. New York: Schocken Books, 1965.

———. *Origins of the Kabbalah*. Princeton, N.J.: The Jewish Publication Society, Princeton University Press, 1987.

Scranton, Laird. *China's Cosmological Prehistory*. Rochester, Vt.: Inner Traditions, 2014.

———. *The Cosmological Origins of Myth and Symbol: From the Dogon and Ancient Egypt to India, Tibet, and China*. Rochester, Vt.: Inner Traditions, 2010.

———. *The Mystery of Skara Brae: Neolithic Scotland and the Origins of Ancient Egypt*. Rochester, Vt.: Inner Traditions, 2016.

———. *Point of Origin: Gobekli Tepe and the Spiritual Matrix for the World's Cosmologies*. Rochester, Vt.: Inner Traditions, 2014.

———. "Revisiting Griaule's Dogon Cosmology: Comparative Cosmology Provides New Evidence to a Controversy." *Anthropology News* 48, no. 4 (April 2007): 24–25.

———. *Sacred Symbols of the Dogon: The Key to Advanced Science in the Ancient Egyptian Hieroglyphs*. Rochester, Vt.: Inner Traditions, 2007.

———. *The Science of the Dogon: Decoding the African Mystery Tradition*. Rochester, Vt.: Inner Traditions, 2006.

Snodgrass, Adrian. *The Matrix and Diamond World Mandalas*. 1988. Reprint, New Delhi: International Academy of Indian Culture and Aditya Prakashan, 1997.

————. *The Symbolism of the Stupa.* Delhi, India: Motilal Banarsidass Publishers, 1992.

Strassler, Matt. "Virtual Particles: What Are They?" Article posted on the Profmattstrassler website.

Thomson, David W., III. *Secrets of the Aether: Unified Force Theory, Dark Matter, and Consciousness.* Alma, Ill.: Quantum Aether Dynamics Institute, 2007.

Tregear, Edward. *The Maori-Polynesian Comparative Dictionary.* Wellington, New Zealand: Lyon and Blair, 1891.

Wilkinson, Toby A. H. *Early Dynastic Egypt.* New York: Routledge, 1999.

Yukteswar, Jnanavatar Swami Sri. *Kaivalya Darsanam: The Holy Science.* 8th ed. Los Angeles: The Self-Realization Fellowship, 1990.

INDEX

BOOKS OF RELATED INTEREST

The Mystery of Skara Brae
Neolithic Scotland and the Origins of Ancient Egypt
by Laird Scranton

Point of Origin
Gobekli Tepe and the Spiritual Matrix for the
World's Cosmologies
by Laird Scranton

Decoding the African Mystery Tradition
by Laird Scranton
Foreword by John Anthony West

Sacred Symbols of the Dogon
The Key to Advanced Science in the
Ancient Egyptian Hieroglyphs
by Laird Scranton
Foreword by John Anthony West

The Velikovsky Heresies
Worlds in Collision and Ancient Catastrophes Revisited
by Laird Scranton

The Great Pyramid Hoax
The Conspiracy to Conceal the
True History of Ancient Egypt
by Scott Creighton
Foreword by Laird Scranton

Black Genesis
The Prehistoric Origins of Ancient Egypt
by Robert Bauval and Thomas Brophy, Ph.D.

Forgotten Civilization
The Role of Solar Outbursts in Our Past and Future
by Robert M. Schoch, Ph.D.

INNER TRADITIONS • BEAR & COMPANY
P.O. Box 388 • Rochester, VT 05767
1-800-246-8648 • www.InnerTraditions.com

Or contact your local bookseller